Framework

MATHS

8 C

HOMEWORK BOOK

David Capewell	Formerly Westfield School, Sheffield
Marguerite Comyns	Queen Mary's High School, Walsall
Gillian Flinton	All Saints Catholic High School, Sheffield
Paul Flinton	Chaucer School, Sheffield
Geoff Fowler	Maths Strategy Manager, Birmingham
Derek Huby	Mathematics Consultant, West Sussex
Peter Johnson	Wellfield High School, Leyland, Lancashire
Penny Jones	Waverley School, Birmingham
Jayne Kranat	Langley Park School for Girls, Bromley
Ian Molyneux	St. Bedes RC High School, Ormskirk
Peter Mullarkey	Netherhall School, Maryport, Cumbria
Nina Patel	Ifield Community College, West Sussex

OXFORD
UNIVERSITY PRESS

OXFORD
UNIVERSITY PRESS

Great Clarendon Street, Oxford OX2 6DP

Oxford University Press is a department of the University of Oxford.
It furthers the University's objective of excellence in research,
scholarship, and education by publishing worldwide in

Oxford New York

Auckland Cape Town Dar es Salaam Hong Kong Karachi
Kuala Lumpur Madrid Melbourne Mexico City Nairobi
New Delhi Shanghai Taipei Toronto

With offices in

Argentina Austria Brazil Chile Czech Republic France Greece
Guatemala Hungary Italy Japan Poland Portugal Singapore
South Korea Switzerland Thailand Turkey Ukraine Vietnam

Oxford is a registered trade mark of Oxford University Press
in the UK and in certain other countries

British Library Cataloguing in Publication Data

Data available

ISBN 978 0 19 914890 5

ISBN 0 19 914890 2

10 9 8 7 6 5 4 3

The photograph on the cover is reproduced courtesy of
Graham Peacock.

The publishers would like to thank QCA for their kind permission to use
Key Stage 3 SAT questions.

Typeset by Bridge Creative Services, Bicester, Oxon

Printed in Great Britain by Bell and Bain, Glasgow

About this book

Framework Maths Year 8C has been written specifically for Year 8 of the Framework for Teaching Mathematics. It is aimed at students who are following the Year 8 teaching programme from the Framework.

The authors are experienced teachers and maths consultants, who have been incorporating the Framework approaches into their teaching for many years and so are well qualified to help you successfully meet the Framework objectives.

The books are made up of units based on the medium-term plans that complement the Framework document, thus maintaining the required pitch, pace and progression.

This Homework Book is written to support the Core objectives in Year 8, and is designed to support the use of the Framework Maths Year 8C Student's Book.

The material is ideal for homework, further work in class and extra practice. It comprises:
◆ A homework for every lesson, with a focus on problem-solving activities.
◆ Worked examples as appropriate, so the book is self-contained.
◆ Past paper SAT questions at the end of each unit, at Level 5 and Level 6 so that you can check students' progress against National Standards.

Problem solving is integrated throughout the material as suggested in the Framework.

Contents

1 The answer to my question is ⁻42.3

My question could be ⁻40 + ⁻2.3.

Make up five different types of questions with the answer of ⁻42.3:

◆ an addition

◆ a subtraction

◆ a word problem

◆ a puzzle

◆ a missing term in a number pattern.

2 Decide whether to use a mental or a written method to calculate:

a	6 + ⁻2	**b**	⁻12 – 3 – 7
c	29 – ⁻39	**d**	139 + ⁻127
e	484 – ⁻209	**f**	473 + ⁻369 + ⁻431
g	⁻926 + ⁻174	**h**	882 + ⁻218
i	⁻937 – ⁻463	**j**	421 – ⁻579

3 In a number pyramid, you add two numbers that are next to each other to get the number above.

Copy and complete these pyramids:

a

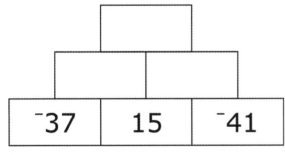

b

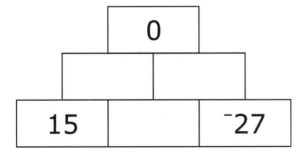

1   Calculate these, deciding whether to use a mental or a written method:

a -6×-5 b $80 \div -20$ c 16×-5 d $-96 \div 6$

e -14×19 f -19×-19 g $294 \div -14$ h -15×18

i -3.6×10 j $-24 \div -1000$

2   a   Answer the questions and match the letters with the answers to solve the puzzle:

N -12×-5

E -8.4×11

C 0.3×-200

D $697 \div 41$

G $-272 \div 16$

O $-924 \div -10$

A $288 \div -18$

What did the Captain say when his ship blew up?

17	-92.4	-60	-16	-17	92.4	60

b   I am thinking of a number. When I multiply it by -17 and subtract -13 the answer is 370.

What number am I thinking of?

3   **Puzzle**

In negative countdown you must use all of the numbers given to make the target number.

You may add, subtract, multiply or divide the numbers.

a   Target number = 75

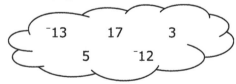

b   Target number = -84

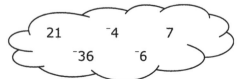

1

> **Remember:**
>
> You can write any number as the product of its prime factors:
>
> $28 = 4 \times 7$
>
> $28 = 2 \times 2 \times 7$

Express each of these numbers as the product of their prime factors.

a 76   **b** 240

c 520   **d** 1280

e 3432   **f** 1955

2 Find the HCF and LCM of each of these sets of numbers:

a 30 and 40

b 210 and 220

c 882 and 1911

d 20, 30 and 45.

> HCF = highest common factor.
> LCM = lowest common multiple.

3 **Investigation**

In this 6 x 5 grid a diagonal cuts through 10 of the squares.

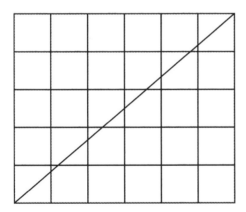

Investigate how many squares the diagonal cuts through in different-sized grids.

Can you find a connection to the HCF and LCM of the grid edges?

1

> **Remember:**
>
> You can write a number as a product of prime factors using index notation:
>
> $25 = 5 \times 5 = 5^2$

Use prime factor decomposition to write each of these numbers using index notation:

a	81	**b**	125	**c**	2187
d	169	**e**	343	**f**	1000
g	1728	**h**	1024		

2 In total countdown you must use all of the numbers given to make the target number.

You may add, subtract, multiply or divide the numbers.

You may also use any number as a power, and you may use the square root sign.

a Target number = 5

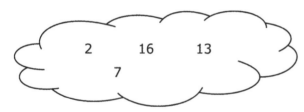

b Target number = 25

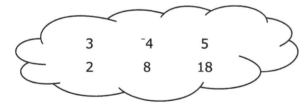

c Invent a countdown question of your own.

3 Find $\sqrt{50}$ by trial and improvement.

Give your answer correct to 1 decimal place.

Remember:
Flow chart boxes

Start
or end

Command

Yes or
No question

1 Follow the instructions to generate a sequence using this flow chart.

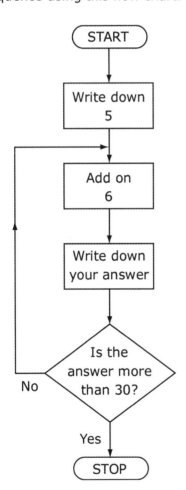

2 Copy this flow chart. Fill in the instructions to generate this sequence:

17, 14, 11, 8, 5, 2, ⁻1

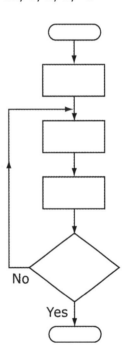

No

Yes

3 The third and fifth terms of this sequence are given.

__ , __ , 4, __ , 16, __ , ...

a Write down the first six terms of **two** different sequences with these third and fifth terms.

b Describe each of your sequences in words.

> **Remember:**
> A linear sequence has a constant difference pattern.
> If the common difference is 2, the nth term is $T(n) = 2n +$ something

1 Write down the first six terms and the general term $T(n)$ of these linear sequences.

    a $T(1)$ and $T(2)$ are 4 and 8

    b $T(2)$ and $T(4)$ are 4 and 8

    c $T(1)$ and $T(3)$ are 4 and 8

    d $T(1)$ and $T(2)$ are 1 and 7

    e $T(1)$ and $T(2)$ are 7 and 12

2 Here are the first three patterns of a sequence.

    a Draw the next pattern in the sequence.

    b Write down the numbers of lines in the first six patterns of the sequence.

    c Predict the number of lines in the 15th pattern. Explain in words how you worked out your answer.

    d Explain how you know that the 100th pattern has 204 lines.

    e Write down the expression for the general term $T(n)$ for this sequence.

3 The general term $T(n)$ for a sequence is $3n - 1$.

    a Write down the first six terms of this sequence.

    b Draw patterns to show the first three terms of this sequence.

    c Using your diagrams, explain why the 100th term in the sequence is 299.

Level 5

You can make 'huts' with matches.

 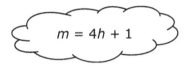

| 1 hut needs 5 matches | 2 huts need 9 matches | 3 huts need 13 matches |

A rule to find how many matches you need is:

$$m = 4h + 1$$

m stands for the number of matches.
h stands for the number of huts.

a **Use the rule** to find how many matches you need to make **8** huts.
Show your working. *2 marks*

b I use **81 matches** to make some huts.
How many huts do I make?
Show your working. *2 marks*

c Andy makes different 'huts' with matches.

  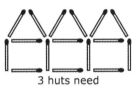

| 1 hut needs 6 matches | 2 huts need 11 matches | 3 huts need 16 matches |

Write down the rule in the cloud that shows how many matches he needs.

Remember: m stands for the number of matches.
h stands for the number of huts.

$$m = h + 5 \qquad m = 4h + 2 \qquad m = 4h + 3$$
$$m = 5h + 1 \qquad m = 5h + 2 \qquad m = h + 13$$

1 mark

Level 6

This is a series of patterns with grey and white tiles.

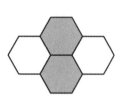

pattern number 1

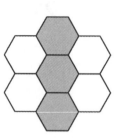

pattern number 2

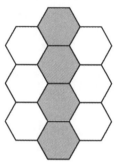

pattern number 3

The series of patterns continues by adding 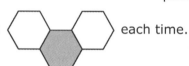 each time.

a   Copy and complete this table:

pattern number	number of grey tiles	number of white tiles
5		
16		

2 marks

b   Copy and complete this table by writing **expressions**:

pattern number	expression for the number of grey tiles	expression for the number of white tiles
n		

2 marks

c   Write an expression to show the **total** number of tiles in pattern number n.
Simplify your expression.

1 mark

1   Find all the labelled angles, explaining your reasons.

Remember:
- Angles on a straight line = 180°.
- Vertically opposite angles are equal.
- Corresponding angles are equal.
- Alternate angles are equal.

a

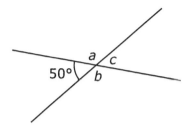

b

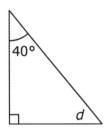

c

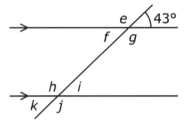

d

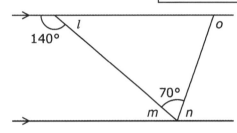

e

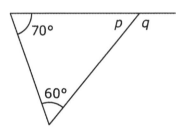

f

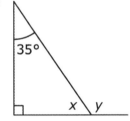

g

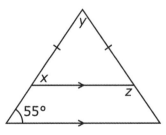

h

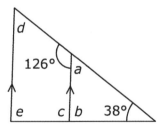

2   Prove that $a = b$.

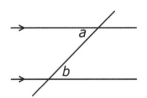

9

1 Find all the labelled angles, explaining your reasons.

a

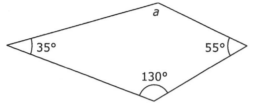

35° | a | 55°
130°

b

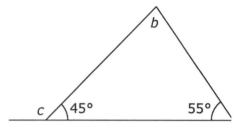

b

c | 45° | 55°

c

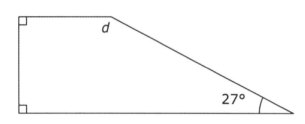

d

27°

d

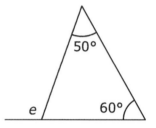

50°

e | 60°

2 Find the angle sum of a hexagon.

3 A quadrilateral has angles 120°, 90°, 60°, x°
Find x and sketch the quadrilateral.

4 A triangle has angles 105°, 60° and y°.
Find y and sketch the triangle.

5 Find z.

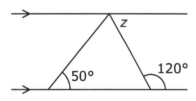

z

50° | 120°

6 Prove that $x = a + b$.

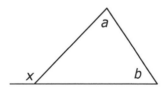

a

x | b

7 Prove that the angle sum of a quadrilateral is 360°.

1   Which of these shapes can you draw?
    Explain carefully the reasons for your answers.

    a   A triangle with one acute angle.
    b   A triangle with two equal angles.
    c   A triangle with three different angles.
    d   A triangle with an obtuse angle.
    e   A triangle with a reflex angle.
    f   A triangle with three lines of symmetry.
    g   A triangle with two lines of symmetry.
    h   A triangle with one line of symmetry.
    i   A triangle with no lines of symmetry.
    j   A triangle with two acute angles.
    k   A triangle with three acute angles.

2   Calculate the angles marked with letters.

a 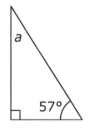   **b** 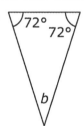   **c**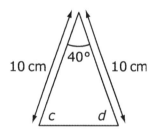

d 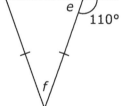   **e**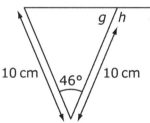

f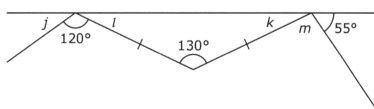

1 Sketch and label two examples of each of these shapes.

Your sketch should include angles, and show any equal sides and lines of symmetry.

a   square

b   rectangle

c   parallelogram

d   trapezium

e   rhombus

f   isosceles trapezium.

2 Calculate the angles marked with letters.

a   **b**

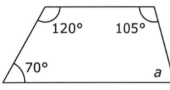

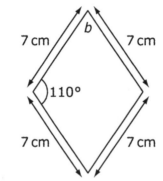

3 On a 3 by 3 grid, find and classify sixteen different quadrilaterals.

For example:

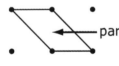parallelogram

square

1   Draw a line that is 12 cm long.

Construct the perpendicular bisector of your line.
Label the midpoint *m*.

2   Draw an obtuse angle and construct the angle bisector.

3   Draw an acute angle and construct the angle bisector.

4   Draw a triangle:

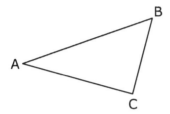

Construct the angle bisector for each of the three angles.
Mark the point where they meet as *m*.

5   Draw a triangle:

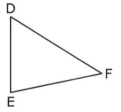

Construct the perpendicular bisector for each of the three sides.

6   Draw two points, A and B.
Draw all the points which are equidistant from A and B.

A•   •B

7   Construct angles of:

    a   60°

    b   45°

    c   30°

> **Hint:**
>
> To make a 60° angle, construct an equilateral triangle with your compasses. For 45° and 30°, bisect 90° and 60° angles.

13

1 Copy each diagram and construct the perpendicular from the point to the line.

a

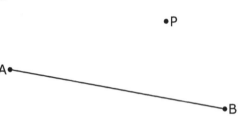

b

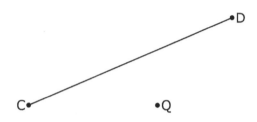

2 Copy each diagram and construct the perpendicular from the points marked on each of these lines.

a

b

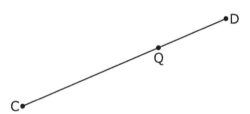

3 Four swimmers are heading for the beach.

Trace this diagram:

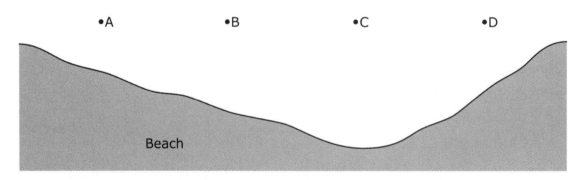

Construct the paths showing the shortest distance to the beach for each of the swimmers.

Julie wants to make a card with a picture of a boat.

The boat will stand up as the card is opened.

Julie makes a rough sketch of the boat.

It is made out of a triangle and a trapezium.

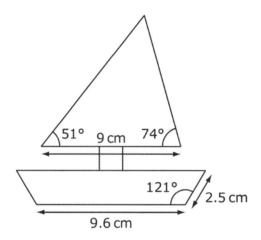

a   On a sheet of paper, make an accurate full size drawing of the **triangle** for the sail.

You need a ruler and an angle measurer or a protractor.   *3 marks*

b   Copy and complete this accurate full size drawing of the **trapezium** for the boat.

You need a ruler and an angle measurer or a protractor.

The bottom of the trapezium, and one edge, have been drawn for you.

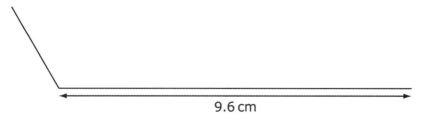

2 marks

Level 6

This shape has **3 identical white** tiles and
3 identical grey tiles.

The sides of each tile are all the same length.

Opposite sides of each tile are parallel.

One of the angles is 70°

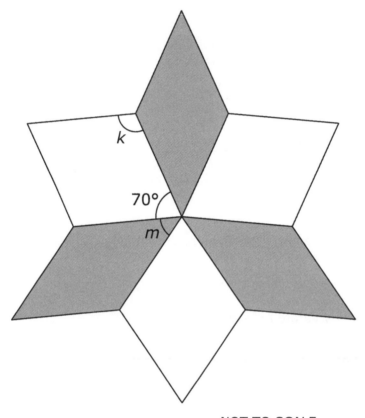

NOT TO SCALE

a Calculate the size of **angle k**.

1 mark

b Calculate the size of **angle m**.
Show your working.

2 marks

> **Remember:**
> Probability = $\dfrac{\text{Number of favourable outcomes}}{\text{Total number of outcomes in sample space}}$

1   List all of the outcomes in the sample space for each of these trials:

   a   A chef checks an egg to see whether it is fresh.

   b   Letter tiles marked M, E, D, I, A and N are put in a bag,
          and one of them is picked out at random.

   c   A boy drops a piece of hot buttered toast on the floor.

   d   Cards marked with the first six multiples of seven are put in a box.
          One card is picked at random.

2   For each of the trials in question **1**, explain whether or not the
        outcomes are equally likely. Explain your reasons.

3   In a game at a school fete, tickets numbered 1–500 are placed in a box.
        Players win a prize if they choose a number that is a
        multiple of 25.

   a   How many outcomes are there in the sample space?

   b   How many favourable outcomes are there?

   c   What is the probability of winning a prize?

4   Write down the name of the month that you were born.

        If you choose a letter at random, what is the probability
        that it is a vowel?

5   An ordinary pack of 52 playing cards has four suits: ♥ ♣ ♦ ♠

        Each suit has 13 cards: number cards from 1 to 10 and 3 picture cards.

        If you choose a card at random from an ordinary pack of 52
        playing cards, what is the probability that the card:

   a   Will be a picture card (Jack, Queen, King)?

   b   Will **not** be a picture card?

   c   Will have a prime number?

   d   Will **not** have a prime number?

1 A red dice is marked 1, 2, 3, 4, 5, 6.

A blue dice is marked 7, 8, 9, 10, 11, 12.

The dice are rolled together, and the scores are added.

Copy and complete the sample space diagram to show all the possible outcomes.

Total score		Score on red dice					
		1	2	3	4	5	6
	7						
	8						
Score on	9					14	
blue dice	10						
	11						
	12						

2 An ordinary dice and a fair spinner marked 1, 2, 3, 4, 5 are rolled and spun together.

The score is the larger of the two numbers showing.

The dice shows 5 and the spinner shows 1. The larger score is 5.

Draw a sample space diagram to show all the possible outcomes.

3 The Galois G9 sports car is available in three engine sizes: 1600 cc, 2000 cc and 2500 cc.

The 1600 cc engine is fitted as standard; the 2000 cc engine costs an extra Euro 600, and the 2500 cc engine is Euro 950 more than the standard unit.

The G9 is also available with three different types of seat: basic, sports (which costs an additional Euro 250) and leather (Euro 500 more than the basic type).

Draw a table to show the total extra cost for all the different combinations of engine size and seat type.

1   A green dice is marked 1, 3, 5, 7, 9, 11.
    A yellow dice is marked 3, 4, 5, 6, 7, 8.

    Copy and complete the sample space diagram to show the total score.

Total score		Score on green dice					
		1	3	5	7	9	11
	3						
	4						
Score on	5					14	
yellow dice	6						
	7						
	8						

Use your diagram to find the probability that the total score is:

a   less than 10   b   more than 20

c   an even number   d   a prime number

e   a multiple of 5.

2   Blaise goes to eat at the Lucky Dip Restaurant. Here is part of the menu.

Starters	Price
1 Snails	£5.50
2 Sardines	£3.75
3 Crab	£4.99
4 Prawns	£4.00
5 Salad	£2.50
6 Mushrooms	£3.50

Main Courses	Price
1 Turbot	£12.50
2 Turkey	£7.50
3 Omelette	£6.00
4 Pigeon Pie	£15.00
5 Beetroot Stew	£9.45
6 Stuffed Peas	£7.75

Blaise rolls an ordinary dice once to choose his starter, and
then rolls again to choose his main course.

a   Draw a sample space diagram to show the total price of
    every combination of starter and main course.

b   Use your diagram to work out the probability that the
    total cost will be more than £15.

When you drop a drawing pin on a table, it can land 'point up' or 'point down'.

Rhona tested two different drawing pins by dropping each one 100 times.
She recorded her results in these tables:

Drawing pin A

down	down	up	down	down	up	down	down	up	down
down	up	up	up	up	down	up	down	down	down
up	down	down	up	up	up	down	down	down	up
down	down	down	down	up	up	down	down	down	up
down	down	up	down	down	down	up	down	up	down
up	up	down	up	up	up	down	up	down	down
down	down	down	down	down	up	down	up	down	down
up	up	down	down	down	up	down	up	up	down
down	down	down	up	down	up	up	down	down	down
down	down	down	down	down	up	up	up	down	down

Drawing pin B

up	up	down	down	down	up	up	up	up	down
up	down	up	down	up	up	up	down	up	up
down	up	up	up	up	up	down	down	up	up
up	up	up	down	down	down	down	up	down	up
down	up	up	down	up	down	up	down	down	down
up	down	up	down	down	up	up	up	up	down
down	up	down	up	up	down	down	up	up	down
down	down	up	down	up	down	down	up	up	up
down	up	up	up	down	down	down	down	up	down
up	up	up	down	up	down	up	up	down	up

1   Use these results to work out the experimental probability of
    each drawing pin landing point down.

2   Write a paragraph to explain what might cause the differences
    in the results for the two drawing pins.

In the television game show 'Spin To Win', the winning contestant is given a prize of £1000 and two coins.

The contestant has to spin the coins:

> If both coins land on heads, £1000 is added to the prize money.

> If you had one head and one tail, £500 is added to the prize money.

> If both coins land on tails, you get half of the prize money, and the game is over.

Provided they did not get two tails, the contestant can then decide whether to keep on spinning, or stop.
If they choose to stop, they keep the prize money.

Two contestants are talking before the show.

> If I win, I'm just spinning once, then stopping.

> I'll just keep going till I get £4000. Then I'll stop.

Test each strategy by spinning coins.

Which strategy do you think is most successful?

This table shows 100 bonus balls from the National Lottery.

36	42	8	24	9	24	37	23	12	9
32	12	49	10	3	13	18	48	13	9
22	14	24	41	3	23	31	6	35	2
8	10	15	39	49	14	11	26	30	42
25	22	1	22	4	34	11	4	1	41
21	17	38	12	38	21	41	16	8	20
16	29	9	38	11	20	28	17	1	46
39	36	17	13	33	3	14	36	6	25
45	40	35	5	16	42	23	40	30	12
3	36	20	43	37	31	42	33	39	14

There are 49 equally likely outcomes for the bonus ball: numbers 1 to 49.

Six of these outcomes (8, 16, 24, 32, 40, 48) are multiples of 8.
The theoretical probability that the bonus ball is a multiple of

8 is $\frac{6}{49}$ = 0.12 to 2 dp.

You can use the data in the table to estimate the experimental probability
that the bonus ball is a multiple of 8.
Using just the first row of the table:

3 of the 10 numbers in the first row are multiples of 8: 8, 24, 24. So the probability is $\frac{3}{10}$.

1   Estimate the experimental probability that the bonus ball is
a multiple of 8, using the first five rows of the table of data.

2   Estimate the same probability using all of the data in the table.

3   Work out the theoretical probability that the bonus ball is a prime number.

4   Estimate the experimental probability of getting a prime number,
using the data in the first row of the table.

5   Work out another estimate of the same probability, using
the data in the first five rows of the table.

6   Use the complete set of data to work out a third estimate for
the probability of the bonus ball being a prime number.

7   Write a short description of your results.
Which estimates agreed most closely with the theoretical probability?

Mark and Kate each buy a family pack of crisps.

Each family pack contains **ten bags** of crisps.

The table shows how many bags of each flavour are in each family pack.

flavour	plain	vinegar	chicken	cheese
number of bags	5	2	2	1

a Mark is going to take a bag of crisps at random from his family pack.

Copy and complete these sentences:

The probability that the flavour will be .... is $\frac{1}{2}$.

1 mark

The probability that the flavour will be **cheese** is ....

1 mark

b Kate ate **two bags** of **plain** crisps from her family pack of 10 bags.

Now she is going to take a bag at random from the bags that are left.

What is the probability that the flavour will be **cheese**?   *1 mark*

c A shop sells **12 bags** of crisps in a large pack.

I am going to take a bag at random from the large pack.

This table shows the probability of getting each flavour.

Use the probabilities to work out **how many bags** of each flavour are in this large pack.

flavour	plain	vinegar	chicken	cheese
probability	$\frac{7}{12}$	$\frac{1}{4}$	$\frac{1}{6}$	0
number of bags				

2 marks

Level 6

There are some cubes in a bag. The cubes are either **red** (R) or **black** (B).

The teacher says:

> If you take a cube at random out of the bag,
> the probability that it will be **red** is $\frac{1}{5}$.

a   What is the probability that the cube will be black?   *1 mark*

b   A pupil takes one cube out of the bag. It is red.

What is the **smallest** number of **black** cubes there
could be in the bag?   *1 mark*

c   Then the pupil takes another cube out of the bag.
It is also red.

From this new information, what is the **smallest** number of
black cubes there could be in the bag?   *1 mark*

d   A different bag has **blue** (B), **green** (G) and **yellow** (Y) cubes
in it. There is at least one of each of the three colours.

The teacher says:

> If you take a cube at random out of the bag,
> the probability that it will be **green** is $\frac{3}{5}$.

There are **20** cubes in the bag.
What is the **greatest** number of yellow cubes there
could be in the bag?
Show your working.   *2 marks*

Use the digits in each cloud to make a fraction and its decimal equivalent.
They have been started for you.

1

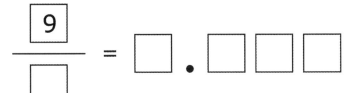

$$\frac{9}{\square} = \square . \square\square\square$$

2

$$\frac{\square}{1\square} = \square . \square\square\overset{\bullet}{3}$$

The dot above the number means it is a recurring decimal – it goes on forever.

3

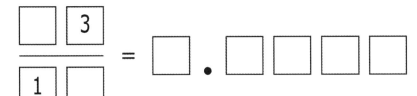

$$\frac{\square 3}{1\square} = \square . \square\square\square\square$$

1   Calculate:

a $\frac{2}{5} + \frac{3}{10}$

b $\frac{3}{4} + \frac{5}{6}$

c $1\frac{2}{9} - \frac{5}{12}$

d $\frac{1}{6} + 0.25 + \frac{1}{8}$

e $6\frac{13}{15} + 1\frac{2}{9}$

f $\frac{32}{33} - \frac{7}{11}$

2   **Investigation**

Jide and Mary are trying to work out:

$\frac{3}{4} + \frac{2}{5}$

Mary shows Jide her working:

$\frac{3}{4} + \frac{2}{5} = \frac{3+2}{4+5} = \frac{5}{9}$

Jide says:

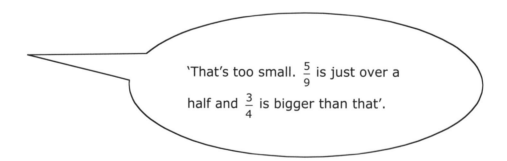

'That's too small. $\frac{5}{9}$ is just over a half and $\frac{3}{4}$ is bigger than that'.

a   What error has Mary made?

b   Work out the correct answer to $\frac{3}{4} + \frac{2}{5}$.

c   Investigate whether the answer will always be too small if you make this error.

1 Calculate these. The first one is done for you.

a $4 \times \frac{2}{5}$

$$4 \times \frac{2}{5} = \frac{4 \times 2}{5} = \frac{8}{5} = 1\frac{3}{5}$$

b $8 \times \frac{5}{7}$ **c** $\frac{4}{9} \times 18$ **d** $\frac{5}{12} \times 48$

e $10 \times \frac{11}{5}$ **f** $18 \times \frac{11}{12}$ **g** $\frac{5}{8} \times 20$

h $3\frac{3}{10} \times 8$ **i** $4\frac{2}{5} \times 3$

2 Calculate these fractions of amounts.

You will need to decide whether to use a mental, written or calculator method.

Give your answers to 2 dp where appropriate.

a $\frac{4}{5}$ of 13   **b** $\frac{11}{12}$ of 108 m

c $\frac{5}{17}$ of £38   **d** $\frac{9}{16}$ of 80 m

e $\frac{8}{11}$ of 250 g   **f** $\frac{3}{8}$ of 84 km

g $2\frac{2}{13}$ of 112 litres   **h** $5\frac{7}{16}$ of 28 tonnes

3 **Puzzle:**

Copy and complete this fraction problem using the digits 0, 2, 5, 7 and 8.

$$\frac{\Box}{\Box} \times £9\Box = £8\Box.\Box0$$

27

Remember:

Use an equivalent fraction:

To find 15% of 40

$$15\% \text{ of } 40 = \frac{15}{100} \times 40$$

$$= \frac{1}{100} \times 15 \times 40$$

$$= \frac{15 \times \cancel{40}^4}{\cancel{100}^{10}} = \frac{60}{10}$$

$$= 6$$

Use an equivalent decimal:

To find 15% of 40

$$15\% \text{ of } 40 = \frac{15}{100} \times 40$$

$$= (15 \div 100) \times 40$$

$$= 0.15 \times 40$$

$$= 6$$

---

1 **a** A cake mixture is 45% flour and 18% sugar.

How much flour is there in 940 g of the mixture?

How much sugar is there in the same amount of mixture?

b On Friday 8% of the students in Year 8 at St Skiver's School were absent.

There are 175 students in Year 8 altogether.

How many students were absent?

c When throwing darts, 'Larger' Len can hit the bullseye 27% of the time.

If Len throws 145 darts, how many times would you expect him to hit the bullseye?

2 Calculate these, giving your answer to 2 dp where appropriate:

a 23% of 64 m   **b** 8% of 320 cm

c 15% of $44   **d** 7% of 1.6 km

e 23.2% of £18 900   **f** 114% of £2.34.

3 **Puzzle**

At the Prontoplant Garden Centre, 45% of the staff go home for lunch.

The remaining 22 staff eat lunch in the canteen.

How many staff work at the garden centre?

1 **a** A jumper costs £18 plus 17.5% VAT.
How much is the jumper?

b In November Carl weighs 86 kg. When he weighs himself
again after Christmas his weight has increased by 9%.

What is Carl's new weight?

c Vicky buys a Zitprone CX2000 sports car for £18,000.

After 1 year the car has gone down in value by 23%.

What is the new value of the car?

d Davina's best ever time to run 800 m is 3 minutes
and 20 seconds.

In the Commonwealth Games she manages to reduce
her personal best for the 800 m by a massive 5%.

What is her new personal best time?

2 Peter Parker spends £3000 on the design of a new internet website.
On the first day of its launch there are 1400 visitors
to the website.

Each day the number of visitors to the website increases by 17%.

a How many visitors are there during the second day?

b How many visitors are there on the fifth day?

c On what day will the number of visitors to the website
top 10 000?

3

Which phone company gives you a better value deal for
the Aztec 300?

1 Mandy and Christabel collect programmes from different events.
Here is a table showing the number of programmes they own.

Type of programme	Mandy	Christabel
Pop concert	14	25
Football match	24	16
Theatre	12	14
Total	50	55

a Who has the greater proportion of pop concert programmes?

b Whose collection is 48% football programmes?

c Who has the higher proportion of theatre programmes?

2 Calculate these. In each case show all your working out and explain your answers.

a Jeanette scores 73% in her History exam and gets $\frac{44}{60}$ in her Geography exam.

In which subject did she do best?

b The probability of scoring a 3 with a single dice is $\frac{1}{6}$.

In an experiment to check if a dice is biased, the results indicate that a 3 was obtained 11.4% of the time.

Is the dice tested a fair dice?

c A receipe for a cake uses 250g of flour, $\frac{1}{4}$ of a pint of milk and two eggs.

What quantities of flour and milk would be needed to make the cake using five eggs?

3 Bernadette pays £2364 in tax out of her yearly earnings of £9850.

Samantha pays £45.14 in tax each week – her weekly wage is £185.

Who pays the highest proportion of tax?

Explain and justify your answer.

Level 5

Hakan asked 30 pupils which subject they liked best.

Subject	Number of boys	Number of girls
Maths	4	7
English	2	4
Science	3	3
History	0	1
French	1	5
	total 10	total 20

a   Which subject did **20%** of **boys** choose?   *1 mark*

b   Which subject did **35%** of **girls** choose?   *1 mark*

c   Hakan said:

'In my survey, **Science** was equally popular with boys and girls'.

Explain why Hakan was **wrong**.   *1 mark*

d   Which subject **was** equally popular with boys and girls?   *1 mark*

Level 6

A report on the number of police officers in 1995 said:

"There were **119 000** police officers. **Almost 15%** of them were **women.**"

a The **percentage** was **rounded** to the nearest whole number, 15.
What is the **smallest** value the percentage could have been,
to one decimal place?
Write down the correct answer from the ones in the bubble.

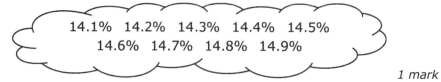

14.1% 14.2% 14.3% 14.4% 14.5%
14.6% 14.7% 14.8% 14.9%

1 mark

b What is the **smallest number** of women police officers that
there might have been in 1995?
(Use your answer to part **a** to help you calculate this answer.)
Show your working. *2 marks*

c A different report gives exact figures:

Number of women police officers	
1988	12 540
1995	17 468

Calculate the **percentage increase** in the number of women
police officers from 1988 to 1995. Show your working. *2 marks*

d The table below shows the **percentage** of police officers in 1995
and 1996 who were women.

1995	14.7%
1996	14.6%

Use the information in the table to decide which one of the
statements below is true. Write down the true statement.

◆ In 1996 there were **more** women police officers than in 1995.

◆ In 1996 there were **fewer** women police officers than in 1995.

◆ There is **not enough information** to tell whether there were
more or fewer women police officers.

Explain your answer. *1 mark*

1 Find the value of these algebraic expressions when $u = 4$, $v = 5$ and $w = 8$. The first one is done for you:

a $uv = 4 \times 5 = 20$ **b** $3v + w$

c $6u - 3v$ **d** $u - 5$

e $vw - 2u$ **f** $\dfrac{4v + w}{2}$

g $2vw - 30$ **h** $3v - w$

i $uw + 6v$ **j** $u - v - w$

2 **Match the batch**

Each equation in box A has a matching equivalent equation in the box B.
Find the four pairs of equations.

A
$3x - 5 = 8$
$3x + 20 = 4$
$4 + 3x = 20$
$8 + 3x = 5$

B
$20 - 3x = 4$
$5 - 3x = 8$
$4 - 3x = 20$
$3x - 8 = 5$

3 In this question, $a = 6$, $b = 2$ and $c = 5$.

Use a, b and c, with any of the signs $+$, $-$, $\times$, $\div$, to make an expression for each value.

The first one is done for you:

a $17 = ab + c$ **b** 13

c 16   **d** 60

e 3   **f** 4

g 28   **h** 7

1   Simplify these algebraic expressions:

   a $4 \times x \times x$
         b $3x \times 5x$

   c $x^2 \times x \times 2$
         d $3x^2 \times 2x^2$

   e $5x^3 \times 3x$
         f $3x + 3x + 4x$

   g $x^5 \times x^3$
         h $3x \times 2x^2 \times 5x^2$

2   Here are eight algebraic expressions.

$x + x + x + y + y$

$2x + 2x + 2y$

$2x + 3x + 5$

$8x + 2y - 4x$

$2y + 3x$

$4y + 2x - 2y + 2x$

$5x - 3$

$5 - 2x + 4x$

   a   Find three expressions that have the same meaning.

   b   Find a pair of expressions that have the same meaning.

   c   Which expression is 8 greater than another expression?

   d   The remaining expression is equal to 10.
       What is the value of x?

3   Work out the value for each of these expressions when $r = 4$
and $s = 5$. The first one is done for you.

   a $2r^2 = 2 \times 4 \times 4 = 16$
      b $2s^2$

   c $r^2 + s^2$
             d $(r + s)^2$

   e $(5r - 4s)^3$
         f $8(s - r)$

   g $(2r - s)^3$
         h $(r - s)^2$

   i $\dfrac{rs}{10}$
             j rs^2

A2.3HW Simplifying expressions

1 In Hot Crosses the sums of the expressions in the horizontal row
and vertical column are the same.
Find the missing expressions in these Hot Crosses.

a

x + 2	3x + 4	2x + 7
	5	

b

	8x − 3	
2x + 2	2x + 5	
	2 − 3x	

c

	3x + 10	
x + 4	x − 8	5x + 9

2 In these towers, the value in each cell is found by adding the
two cells above it.
Copy and complete the addition towers; using only these
expressions. You may use them more than once.

6 − 9x	2x + 9	8x + 5	3x − 2	4 − 6x	11 − x

5x + 7	10x + 14	9x − 6

a

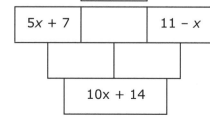

8x + 5	4 − 6x	9x − 6

b

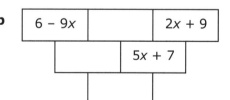

6 − 9x		2x + 9
	5x + 7	

c

5x + 7		11 − x
	10x + 14	

1 Multiply out these brackets.

    a $5(2t + 3)$ **b** $3(6t - 4)$

    c $4(9 - 4t)$ **d** $3(7t + 3s)$

    e $6(2t - 3s)$ **f** $5(t - 8)$

    g $t(t + 3)$ **h** $7(4t - 5s)$

    i $2t(2t + 3)$ **j** $s(3t - 2s)$

2 Copy this diagram.

Add five different ways of writing $18x + 72$.

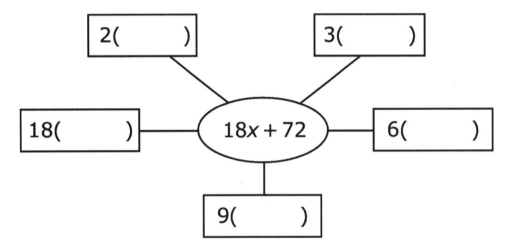

3 Match equivalent expressions from each box.
The first one is done for you.

$6(3x - 2y) = 18x - 12y$

6(3x − 2y)	2(12x − 10y)
8x + 6y	18x − 12y
4(6x − 5y)	4(6x − 2y)
9x − 12y	14x + 6y
2(7x + 3y)	2(4x + 3y)
8(3x − y)	3(10x + 15y)
5(6x + 9y)	3(3x − 4y)

1 Work out the value of each expression when
$a = 3$, $b = {}^-4$ and $c = {}^-1$.

a $3(a + b)$ **b** $2b + 4a$

c $a - b + 3c$ **d** $2b + ac$

e $\dfrac{a + 3b}{3c}$ **f** $ab - c$

g $\dfrac{8a + 3b}{a - c}$ **h** $a^2 - b^2 - b$

Look at your solutions.
You should have four pairs of equal values.
Write down each pair of equal expressions as an equation.

> **Hint:**
> Use an = sign to link two expressions with the same value.

2 In this question, $a = 5$, $b = {}^-2$ and $c = {}^-3$.
Use a, b and c, with any of the signs $+$, $-$ $\times$, $\div$ to make an expression for each value.
You must use at least two of the letters a, b, c in each expression.

The first one is done for you:

a $23 = a^2 + b$ **b** 27

c ${}^-5$ **d** 7

e 0 **f** ${}^-7$

g ${}^-10$ **h** ${}^-1$

i $+3$ **j** 20

The formula for the area (*A*) of a trapezium is given by:

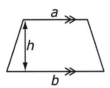

$$A = \frac{1}{2}(a + b)h$$

1 Use the formula to find the area of each trapezium:

a
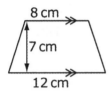
8 cm
7 cm
12 cm

b
4 cm
4 cm
7 cm

c
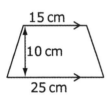
15 cm
10 cm
25 cm

d
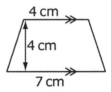
12 cm
8 cm
15 cm

2 This trapezium has an area of 36 cm².
Use the formula to find its height.

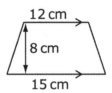
6 cm
h
12 cm

3 This trapezium has an area of 60 cm².
Find possible values for *a*, *b* and *h* for
this trapezium.

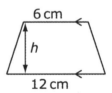
a
h
b

4 **a** Write down the formula for the area of each shape.

b Find values for each of the lengths so that each shape has
area 42 cm².

i
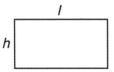
l
h

ii

h
b

iii
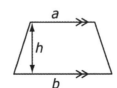
a
h
b

iv
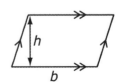
h
b

Level 5

Ali, Barry and Cindy each have a bag of counters.

They do not know how many counters are in each bag.

They know that:

   Barry has **two more** counters than Ali.

   Cindy has **four times as many** counters as Ali.

a   Ali calls the number of counters in her bag **a**.

   Write **expressions using a** to show the number of counters
   in Barry's bag and in Cindy's bag.

Ali's bag   Barry's bag   Cindy's bag

1 mark

b   Barry calls the number of counters in his bag **b**.

   Write **expressions using b** to show the number of counters
   in Ali's bag and in Cindy's bag.

Ali's bag   Barry's bag   Cindy's bag

2 marks

c   Cindy calls the number of counters in her bag **c**.

Ali's bag   Barry's bag   Cindy's bag

   Which of the expressions below shows the number of counters
   in **Barry's** bag?

   Write down the correct one.

   $4c + 2$ $4c - 2$ $\dfrac{c}{4} + 2$

   $\dfrac{c}{4} - 2$ $\dfrac{c + 2}{4}$ $\dfrac{c - 2}{4}$ *1 mark*

Level 6

a   Write an expression for each missing length in these rectangles.
Write each expression as simply as possible.

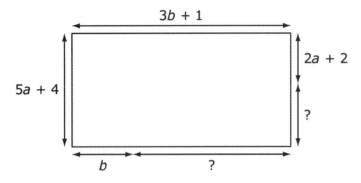

2 marks

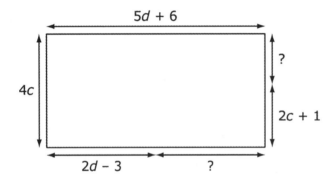

2 marks

b   The length of one side of a rectangle is y.

This equation shows the area of the rectangle:

$$y(y + 2) = 67.89$$

Find the value of y.

Show your working.

You may find the following table helpful.

y	$y + 2$	$y(y + 2)$	
8	10	80	too large

2 marks

1   What units would you use to accurately measure the mass of one piece of paper?

2   To make a pair of jeans Janine uses 227 cm of denim.

How many pairs can she make from a 300 m roll of denim?

3   In Aldo a two litre bottle of milk costs 85p.

A 500 ml bottle costs 22p.

Which is the best value?

Give reasons for your answer.

4   Tables measure 138 cm by 78 cm:

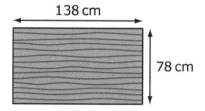

Kit is organising the judging of a hat competition and needs to fit as many tables as possible into a hall measuring 5 m by 7 m.

To allow the judges to move between the tables she must allow 0.5 m between each row of tables and a corridor of 1 m down the centre of the room.

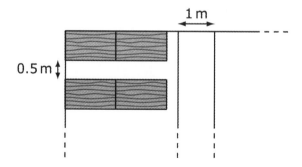

Work out how many tables she can fit in the hall.

Remember these rough equivalents:

Length	Mass	Capacity
1 km ≈ 1.6 miles	1 kg ≈ 2 pounds	1 litre ≈ 2 pints
1 mile ≈ $\frac{5}{8}$ km	1 pound ≈ $\frac{1}{2}$ kg	1 pint ≈ $\frac{1}{2}$ litre

1 The supermarket sells apples for £1.50 per kilogram.

The corner shop sells apples for 58p per pound.

Which is the cheapest? How can you tell?

2 A three litre bottle of coke in a supermarket costs £1.59.

A half pint glass in a café costs 60p.

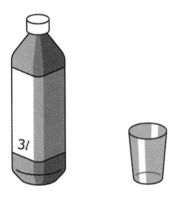

Which is the best value and why?

3 On this signpost, the distance to Rotherham is given as 9 miles.

Rotherham	9	Chesterfield	13
Airport	$\frac{3}{4}$	Norton	$2\frac{1}{2}$

a Do you think this distance is correct to the nearest yard, the nearest quarter mile or the nearest mile?

b Copy the sign and show the distances in kilometres.

1   On squared paper:

  a   Draw ten different triangles with an area of 12 cm^2.

  b   Draw eight different parallelograms with an area of 20 cm^2.

  c   Draw six different trapeziums with an area of 20 cm^2.

2

> **Remember:**
>
> Area of triangle $= \frac{1}{2}$ x base x perpendicular height.
>
> Area of parallelogram = base x perpendicular height.
>
> Area of a trapezium $= \frac{1}{2} (a + b)$ x perpendicular height,
>
> where a and b are the lengths of the parallel sides.

Name each of these shapes, giving a reason for your choice,
then calculate their area.

a

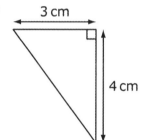

3 cm
4 cm

b

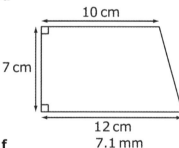

6 cm
5.2 cm

c

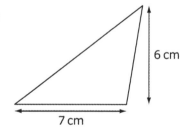

6 cm
7 cm

d

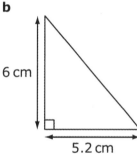

10 cm
7 cm
12 cm

e

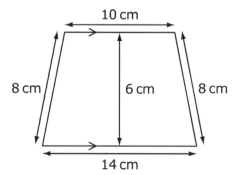

10 cm
8 cm   6 cm   8 cm
14 cm

f

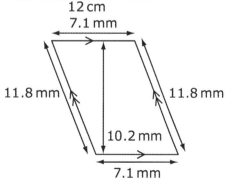

7.1 mm
11.8 mm   11.8 mm
10.2 mm
7.1 mm

1   The diagram shows a shaded square inside a square:

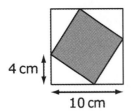

4 cm

10 cm

Calculate the area of the shaded square.

2   **a**   A triangle has an area of 12 cm^2.

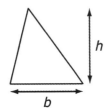

h

b

Write down possible values of b and h.

b   A parallelogram has an area of 16 cm^2

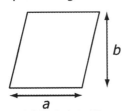

b

a

Write down possible values of a and b.

c   A trapezium has an area of 20 cm^2

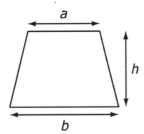

a

h

b

Write down possible values of a, b and h.

1 Calculate the volume of these cuboids.

> **Remember:**
> Volume of a cuboid = length x width x height

a

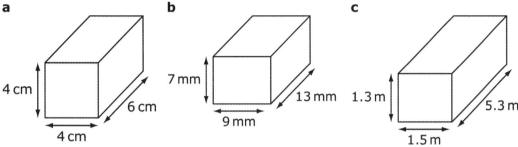

4 cm

6 cm

4 cm

b

7 mm

13 mm

9 mm

c

1.3 m

5.3 m

1.5 m

2 Calculate the volume of this T-shaped girder.

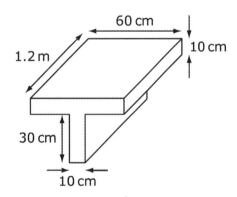

60 cm

10 cm

1.2 m

30 cm

10 cm

3 Boxes of Mint Krisp measure 11 cm by 20 cm by 15 cm. The boxes must stay this way up!

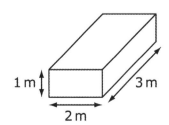

11 cm

20 cm

15 cm

a How many boxes of Mint Krisp will fit in this crate?

b Change the dimensions of the crate to reduce any wasted space.

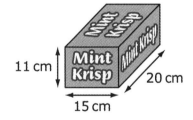

1 m

3 m

2 m

> **Remember:**
> Volume of a cuboid = length x width x height
> Surface area of a cuboid = the total area of its faces

1 Two equal-sized cushions fit together to make a cube of edge 80 cm.

Calculate the volume and surface area of one cushion.

You may find it helpful to draw side and plan views of one cushion.

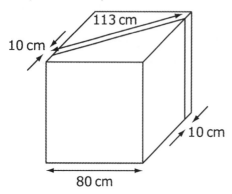

2 A cube has a volume of 216 cm^3.

Find the surface area.

3 Find the surface area and volume of this cuboid:

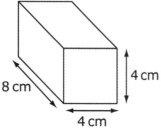

A cube of width, length and height 2 cm is removed
from the centre of the cuboid.
Find the new volume.

4 **Investigation**

A packaging firm wants to produce a cuboid that has the
same volume and surface area.

For example, if the volume is 216 cm^3, then the surface
area must be 216 cm^2.

Find the possible dimensions of the cuboid.

a The diagram shows a rectangle **18 cm** long and **14 cm** wide.

It has been split into **four smaller rectangles**.

Write down the area of each **small rectangle**.

One has been done for you.

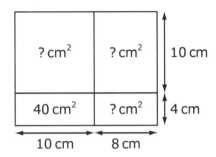

What is the area of the **whole** rectangle?

1 mark

What is **18 × 14**?   *1 mark*

b The diagram shows a rectangle (**n + 3**) cm long and
(**n + 2**) cm wide.

It has been split into **four smaller rectangles**.

Write down a **number** or an **expression** for the **area** of **each
small rectangle**.

One has been done for you.

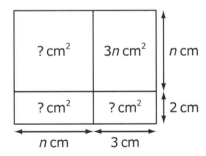

1 mark

Level 6

Each shape in this question has an **area** of **10** cm^2.

a Calculate the height of the parallelogram. area = 10 cm^2

area = 10 cm^2

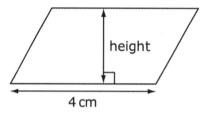

height

4 cm

1 mark

b Calculate the length of the base of the triangle.

area = 10 cm^2

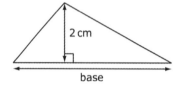

2 cm

base

1 mark

c What might be the values of *h*, *a* and *b* in this trapezium?
(*a* is greater than *b*)

area = 10 cm^2

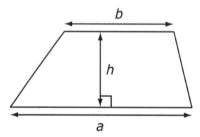

b

h

a

1 mark

What else might the values of *h*, *a* and *b* be?   *1 mark*

d Look at this rectangle:

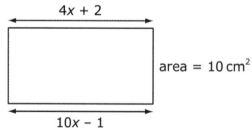

4*x* + 2

area = 10 cm^2

10*x* − 1

Calculate the value of *x* and use it to find the length and
width of the rectangle. Show your working.   *2 marks*

1 Complete each function machine and match it with one of these functions in the box.

a
input output
x y
1
2 → -2 → $\times 2$ →
3

b input output
x y
1
2 → $\times 4$ → -3 →
3

$y = 4x - 3$

$y = \dfrac{x}{2} + 4$

$y = 2(x - 2)$

$y = \dfrac{x + 5}{2}$

c
input output
x y
1
2 → $\div 2$ → $+4$ →
3

d input output
x y
1
2 → $+5$ → $\div 2$ →
3

2 For each of these function machines:
◆ complete the outputs and inputs
◆ write down the rule using algebra.

a
input output
1 ?
2 → $\times 3$ → -5 → ?
x y

b input output
3 ?
? → $+4$ → $\times 5$ → 45
x y

c
input output
? $^-1$
12 → $\div 2$ → -3 → ?
x y

d input output
? $^-6$
5 → $^-4$ → $\times 3$ → ?
x y

3 Draw a function machine for each of these algebraic functions:

a $y = \dfrac{x}{3} + 5$

b $y = \dfrac{x + 5}{3}$

input output

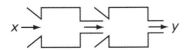
$x \rightarrow$ → y

49

1   For the function $y = 3x - 2$:

   a   Copy and complete the function machine.

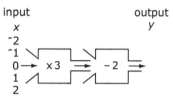

input
x
⁻2
⁻1
0
1
2

output
y

×3   − 2

   b   write down the coordinate pairs made.   Coordinate pairs

(,   )

(,   )

(,   )

(,   )

(,   )

(x , y)

   c   Draw a coordinate grid with x from ⁻2 to 2 and y from
        ⁻10 to 5, and plot these points.

        Join the points with a straight line to represent the linear
        function.

2   For the function $y = 4x - 3$:

   ◆   Copy and complete the table of values,

   ◆   write out the coordinates pairs,

   ◆   plot the coordinates pairs on a grid, and

   ◆   join up with a straight line to draw the line of the equation.

x	⁻2	⁻1	0	1	2	3	
$y = 4x - 5$	⁻13			⁻1			

   Coordinate pairs

(,   )

(,   )

(,   )

(,   )

(,   )

(x , y)

Remember:
◆   A graph with an equation of the form $y = mx + c$:
  ◆   is a straight-line graph
  ◆   crosses the y-axis at $(0, c)$
  ◆   gets steeper as m gets bigger.

---

1   Match each graph on the grid to an equation in the box.

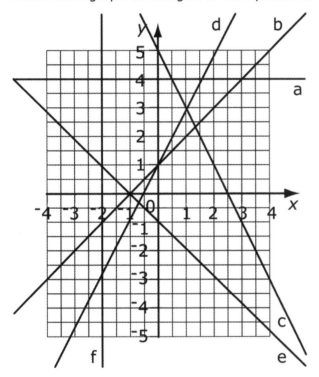

$y = 5 - 2x$	$x = ^-2$
$y = ^-x - 1$	$y = 4$
$y = x + 1$	$y = 2x + 1$

2   Draw a coordinate grid with x from $^-4$ to 4 and y from $^-10$ to 10
for each part of the question.
Sketch on the graphs.
What shape have you made between all of your lines?

  a $y = 3, y = ^-1, x = 4, x = ^-2$

  b $x = 1, x = ^-3, y = x + 1, y = x - 1$

  c $y = ^-1, y = 3, y = 2x + 5, y = 2x - 5$

  d $y = 2, y = ^-3, y = x + 2, y = 4 - x$

1 This graph shows the time taken to oven cook chickens of different weights.

a Copy this chart carefully on to squared paper.

b Use the conversion chart to find how long these chickens should cook.

  i 3 kg   **ii** 3.5 kg

  iii 1.5 kg   **iv** 1 kg

c On Sunday I cooked a chicken for 2 hours 20 minutes. What weight was the chicken?

d The instructions to cook a chicken say "Cook for 50 minutes per kg + 20 minutes".

Explain how this is shown on the graph.

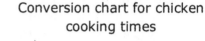
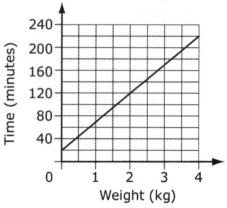

Conversion chart for chicken cooking times

2 Peak rate pay-as-you-go mobile phone charges are very high.

a Copy the grid and draw a conversion chart to show the charges. Use the fact that a 30 minute call at peak time costs £4.50. Start at (0, 0).

b Work out the cost of these peak time calls:

  i 10 minutes   **ii** 13 minutes

  iii 25 minutes   **iv** 50 minutes

c One call cost exactly £1.00.

How long was the call?

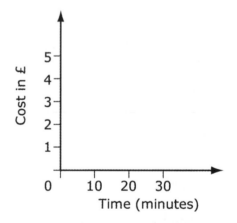

1   This distance–time graph represents Jade's journey to school in the morning.

    a   What time does she leave home?

    b   How long does she wait for the bus?

    c   After getting off the bus she buys a drink at the local shop. How long did this add to her journey time?

    d   How long did it take Jade to get to school?

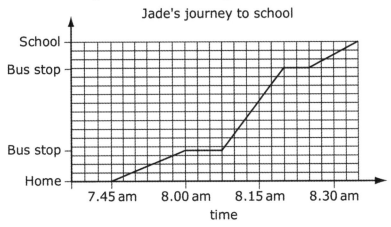

Jade's journey to school

2

The timetable for a train is:

Birmingham	Coventry	Milton Keynes	London Euston
depart 0918	arrive 0936	arrive 1011	arrive 1045
	depart 0940	depart 1013	

Plot these points on a copy of the graph and draw the distance–time graph.

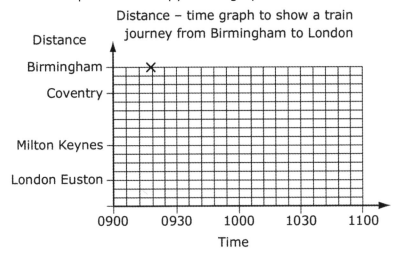

Distance – time graph to show a train journey from Birmingham to London

1 This graph shows the average daily temperature each month for one year in Manchester.

a Which was the coldest month? What was the average temperature that month?

b What was the maximum average daily temperature?

c For what periods was the average daily temperature below 15 °C?

d Grass grows when the average daily temperature is above 12 °C. For how many months did the grass grow?

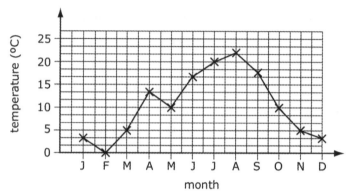

2 This graph shows the depth of water in a children's paddling pool.

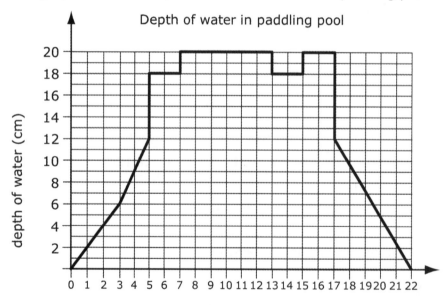

a At first the paddling pool was filling with hot water. When was the cold tap turned on as well?

b Two children got into the padding pool at different times. At what times did they get in?

c Explain what each stage of the graph shows.

Level 5

The grid shows the first eight lines of a spiral pattern.

The spiral pattern starts at the point marked ■.

a   Copy and continue the spiral by drawing the **next four lines** on a square grid.

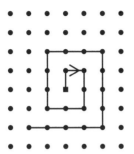

1 mark

b   The table shows the length of each line.

line number	length
1	1
2	1
3	2
4	2
5	3
6	3
7	4
8	4
9	5

The rule for finding the length of **odd** numbered lines is:

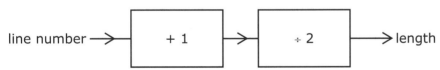

What is the length of line number **23**?   *1 mark*

c   Copy this diagram and fill in the box to show the rule for finding the length of **even** numbered lines.

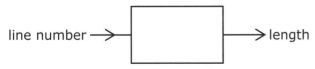

1 mark

d   What is the length of line number **18**?   *1 mark*

Level 6

Look at this diagram:

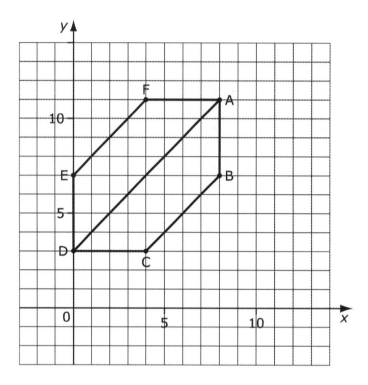

a The line through points A and F has the equation $y = 11$.

What is the equation of the line through points **A** and **B**?

1 mark

b The line through points A and D has the equation $y = x + 3$.

What is the equation of the line through points **F** and **E**?

1 mark

c What is the equation of the line through points **B** and **C**?

1 mark

1 Write down the value of these numbers.

Give your answers in figures and words.

The first one is done for you.

a 2×10^3

> **a** 2×10^3
>
> $2 \times 10^3 = 2 \times 10 \times 10 \times 10$
>
> $= 2 \times 1000$
>
> $= 2000$

b 6.2×10^4

c $^-14.56 \times 10^2$

d 7.002×10^6

2 Write down the answers to these calculations:

a 8×0.1 **b** 0.01×3

c 10×4.7 **d** $6 \div 0.1$

e $19 \div 0.01$ **f** 5.4×0.1

g 168×0.01 **h** $^-3.5 \div 0.01$

i $0.5 \div 0.1$ **j** 0.8×0.01

3 Place the answers to these questions in order from lowest to highest:

a 0.004×10^2

b 42×0.01

c $0.0395 \div 0.1$

d $4.1 \div 100$

e 3.9×0.1

4 Find the number that lies exactly halfway between each of these pairs of numbers:

a 3.4 and 3.9 **b** 0.27 and 0.75

c $^-3.6$ and $^-1.25$ **d** $^-2.64$ and 1.08.

5 Copy and complete these statements:

a $0.35 \times ? = 35$ **b** $0.35 \times ? = 700$

c $? \div 0.1 = 28$ **d** $5.8 \times ? = 0.0058$

e $^-3.7 \div ? = {}^-370$ **f** $^-0.07 \div ? = {}^-7$

1 Round each of these numbers:

    a to the nearest whole number

    b to one decimal place

    c to two decimal places.

    i 7.385

    ii 0.748

    iii 899.189

    iv 12.0692

2 Here are the top four distances thrown by the world's best shot-putters:

Sumo	0.02375 km
Giant	23.746 m
M'sive	2378.8 cm
Lairge	237 700 mm

    a Round each athlete's distance to the nearest unit.

    b Use your results to rank the competitors.

    c Is this a fair way of identifying the best shot-putter?
Explain your answer.

3 Use your calculator to work out these, giving your answers correct to 2 dp.

    a 39% of 27.6 m

    b $\frac{3}{11}$ of 24 kg

    c the product of 25.7, 5.32 and 4.89

    d the quotient when 23 is divided by 24.

4 **Puzzle**

Ruby raises a number (x) to the power of 4. The answer is 501 (correct to the nearest whole number),

i.e. $x^4 = 501$.

What number did she start with? (give your answer correct to 2 dp.)

1   Calculate these using a mental or written method:
The first one is done for you.

    a   ⁻3.3 + 5.3

                              a   ⁻3.3 + 5.3

                                     ⁻3.3 + 5.3 = 5.3 – 3.3

                                                 = 2

    b   2.8 - ⁻1.9 – 2.8   **c**   26.74 + 3.8 – 15.81

    d   16.4 + 24.6 – 23.84   **e**   0.71 – 12.7 + 18.4

    f   56.7 + 213.2 + ⁻99.99   **g**   0.381 km – 0.93 km

    h   2.08 mm – 1.69 mm + ⁻0.8 mm   **i**   1.03 kg + ⁻0.97 kg

    j   33.03 + 3.3 + ⁻3 + 0.33

2   Calculate the perimeter of this letter E.

    Give your answer in metres.

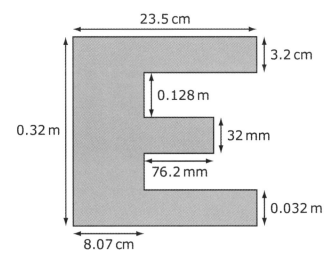

3   The answer to my question is ⁻22.3 (to 1 dp).

    Make up five different types of questions with this answer.

    For example:

    ◆   an addition

    ◆   a subtraction

    ◆   a word problem

    ◆   a puzzle

    ◆   a missing term in a number pattern.

1 Calculate these. You may need to make some jottings.
The first one is done for you.

a   16 x 0.9

a   16 x 0.9

$$16 \times 0.9 = (16 \times 10) - (16 \times 0.1)$$
$$= 160 - 1.6$$
$$= 158.4$$

b ⁻2.7 x 31   **c**   4.5 x 8   **d**   216 ÷ 24

e   4 x 0.09   **f**   ⁻3.4 ÷ 0.1   **g**   1.5 x 21

h   3.7 x 19   **i**   ⁻25 x 12   **j**   ⁻288 ÷ ⁻12

k   4.7 x 13   **l**   ⁻3.8 x 1.5

2 **Puzzle**

a Copy and complete this multiplication grid:

x	30		
	600	140	
		35	1.5
	6		0.06

b Use your grid to work out 25.2 x 37.3.

3 **Puzzle**

Find the correct digit to go in each of these spaces.
Some of the questions have more than one solution.

a   ___ 6 x 1 ___ = 4 ___ 8

b   17 x ___.3 = ___ 3.1

c   ⁻___ .2 x ___ 3 = ⁻___ 3.6

d   ⁻4 ___ ___ ÷ 18 = ⁻___ 3

1 Calculate these, using a mental or written method as appropriate:

   a 0.4 x 0.9   **b** 3.47 x 4

   c 7.23 x 8   **d** 15.8 x 0.6

   e 21.7 x 2.3   **f** 18.2 x 1.5

   g 3.22 x 25   **h** 29 x 8.3

   i 28 x £3.71   **j** 19.6 m x 2.8 m

   k 17.9 km x 8.9

2   **a** A bag of rice from the supermarket costs £1.20 a kilogram.
Siobhan needs 13.8 kilograms of rice.

      How much will the rice cost Siobhan?

   b A box of 48 packets of crisps weighs 2.64 kg.

      What is the weight of 12 boxes?

   c The average height of a boy in Year 8 is 1.64 m.

      There are 96 boys in Year 8.

      If all the boys in Year 8 lay on the ground head to toe,
how long a line would they make?

3   **a** Copy and complete the table of values from $x = {}^-3$ to $x = 3$
for the equation $y = 3.7x - 0.6$.

x	$^-3$	$^-2$	$^-1$	0	1	2	3
$y = 3.7x - 0.6$							

   b Draw the line $y = 3.7x - 0.6$ on a grid with x-axis from $^-3$ to 3
and y-axis from $^-35$ to 35.

   c Does the point (5, 18) lie above or below the line
$y = 3.7x - 0.6$? Explain your answer.

4 **Puzzle**

Copy and complete these multiplications by filling in the missing digits.

   a 34. ____ x ____.3 = 2____3.86

   b ____7.9 x 4____ = 1____9____.7

N3.6HW   Dividing decimals

1 Copy and complete this bill from McVeggies Fast Food.

Item	Cost per item	Number of items	Total cost
Beanburger	£1.96	8	
Veggie meal		4	£15.48
Fries		9	£10.98
Baked potato		3	£4.77
Pop (0.75 litre)		7	£5.95
Juice (0.4 litre)		5	£3.95
		Total	

2 Calculate:

a 134.4 ÷ 16   **b** 79.4 ÷ 17

c 174.2 ÷ 26   **d** 284.2 ÷ 58

e 300 ÷ 57   **f** 128 ÷ 37

g 172.8 m ÷ 32   **h** 203.2 ÷ 32

3 **a** How much money will each person receive if 19 friends share £438?

    b A box weighs 3.24 kg when empty and 14.94 kg when filled with 24 tins of carrots.

        What is the weight of one tin of carrots?

    c Gina runs as fast as she can for 30 seconds. She covers a total distance of 222.6 metres.

        How far does she run in one second?

        How far would she run in 100 seconds?

4 Use the digits 2, 4, 5, 6 and 7 to complete this division.

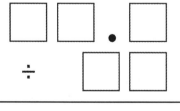

1 Express each of these as ratios in their simplest form.
The first one is done for you.

a   5 m : 140 cm   **a**   5 m : 140 cm

5 m : 140 cm = 1 m : $\frac{140}{5}$ cm

= 1 m : 28 cm

b   42 : 35 : 63

c   3.5 kg : 1400 g

d   350 mm : 2.25 m : 75 cm

2 Express the following as ratios in their simplest form.

a   A recipe requires 250 g of breadcrumbs for every 0.4 kg of margarine.
What is the ratio of breadcrumbs to margarine?

b   In a week, Bernhard earns £4.60 per hour for 24 hours;
Colin earns £6.40 per hour for 18 hours; and Padhrig
earns £5.80 per hour for 22 hours.

What is the ratio of:
Bernhard's weekly wage : Colin's weekly wage : Padhrig's weekly wage?

c   Damian is 1.84 m tall. Sian is 23 cm shorter in height.
What is the ratio of Damian's height : Sian's height?

3 **a**   A 12.8 m length of piping is divided into two pieces in the
ratio 9 : 7. How long is each of the two pieces?

b   A restaurant bill of £78.40 is shared between Mark, Karl
and Bernard in the ratio 10 : 12 : 13.

How much does each person have to pay?

4 **a**   An alloy is made from cobalt and chromium in the ratio 7 : 5.
How much cobalt is needed to mix with 345 g of chromium?

b   The scale of a map is 1 : 10 000.

How long is a distance of 2.35 km in real life on the map?

c   Charmaine saves $\frac{3}{5}$ of the amount she spends.

If she saves £121.32 in a week, how much did she spend?

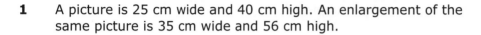

1   A picture is 25 cm wide and 40 cm high. An enlargement of the same picture is 35 cm wide and 56 cm high.

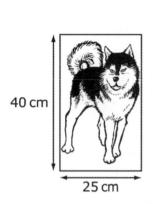

40 cm

25 cm

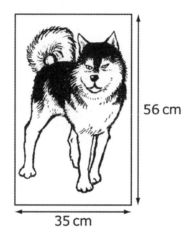

56 cm

35 cm

a   How many times wider is the enlargement than the original picture?

b   What fraction of the height of the enlargement is the height of the original?

c   What is the ratio of the width of the original picture to the width of the enlargement?

d   Write four more statements about the heights and widths of the original picture and the enlargement.

2   **a**   Helen is 80 kg in weight.

Laura weighs 64 kg.

i   How many times heavier is Helen than Laura?

ii   What fraction of Helen's weight is Laura's weight?

iii   What is the ratio of Helen's weight to Laura's weight?

b   Donna spends £7800 on a new car.

Three years later her car is valued at £5070.

i   By what scale factor has the value of her car decreased?

ii   How many times more valuable was her car 3 years ago?

iii   What is the ratio of the original value of the car to the new value of the car?

> **Remember:**
>
> Scale factor = $\dfrac{\text{new value}}{\text{original value}}$

c   Make up a similar word problem on ratios of your own.

1   The scale factor for changing inches into centimetres is $\frac{5}{2}$.

$$\times \frac{5}{2}$$

| number of inches | → | number of centimetres |

a   Use this information to copy and complete this conversion table.

Inches	5	24		
Centimetres	10	20		

b   What is the scale factor for changing the number of centimetres into the number of inches?

2   The numbers in each of these tables are proportional.
In each case, find the missing numbers.

a

Number of packets	Cost (p)
5	235
13	

b

Number of tins	Cost (£)
7	3.71
	12.72

3   Solve each of these problems.
In each case, clearly show the method you have used.

a   15 litres of petrol cost £10.80.
How much would 30 litres of petrol cost?

b   12 apples cost £3.72. How many apples could you buy for £15?
How much money would you have left?

c   9 pizzas cost £17.91. How much would 5 pizzas cost?

d   A distance of 5.2 km is represented by 13 cm on my map.
How many cm will represent 1.3 km?

Level 5

a Robert and Gwen must put 63 tins of food into a lift.

Each tin weighs **840 g**.

Work out the total weight of the **63** tins in grams.

Remember to write down enough working to show you have not used a calculator.   *2 marks*

b In the lift there is a sign.

It shows the greatest load that the lift can carry safely.

Look at the total weight of the 63 tins, which you worked out in part **a**.

Is it safe to carry the 63 tins together in the lift?

Give a reason for your answer.   *1 mark*

> **Greatest load 50 kg**

c Robert and Gwen must put the tins into a cupboard.

All the tins are the same size.

Gwen has measured the height of a **tin**. It is **14cm.**

Robert has measured the height of the **cupboard**. It is **1.24m**.

How many **layers** of tins can they keep in the cupboard?

Remember to write down enough working to show you have not used a calculator.   *3 marks*

Level 6

A company sells and processes films of two different sizes.

The tables show how much the company charges.

Film size: **24** photos	
Cost to **buy** each film	£2.15
Postage	free
Cost to **print** each film	£0.99
Postage for each film	60p

Film size: **36** photos	
Cost to **buy** each film	£2.65
Postage	free
Cost to **print** each film	£2.89
Postage for each film	60p

I want to take **360** photos.

I need to buy the film, pay for the film to be printed, and pay for the postage.

Is it cheaper to use all films of size 24 photos, or all films of size 36 photos?

How much cheaper is it? Show your working.   *4 marks*

> **Remember:**
> Congruent shapes are the same shape and size.

1   Two congruent right-angled triangles can be fitted together to make a rectangle.

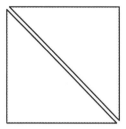

What other shapes can you make with two congruent right-angled triangles?

Draw sketches to illustrate your answers.

2   Two congruent isosceles triangles can be fitted together to make a rhombus.

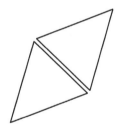

What other shapes can you make?

Draw sketches to illustrate your answers.

3   Investigate the different shapes you can make with two congruent scalene triangles.

Here is a shape on a grid:

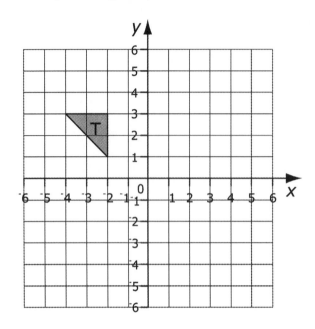

Copy the grid and shape, then answer these questions.

1 **a** Draw the image of T after a translation through the vector $\binom{-1}{3}$.
Label the image T′.

  b What translation will map T′ back to T?

2 **a** Draw the image of T after a reflection in the *y*-axis.
Label the image T″.

  b What reflection will map T″ back to T?

3 **a** Draw the image of T after a rotation through 90° clockwise
about (0, 0).
Label the image T‴.

  b What rotation will map T‴ back to T?

4 **a** What transformation will map T‴ onto T″?

  b What transformation will map T‴ onto T′?

1   Copy this shape and grid:

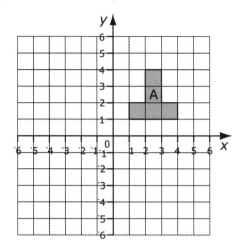

a   Reflect A in the *y*-axis and label the image A′.

b   Reflect A′ in the *x*-axis and label the image A″.

c   What single transformation is equivalent to a reflection in the
    y-axis followed by a reflection in the *x*-axis?

2   Repeat question **1**, but reverse the order of the transformation:
    reflect A in the *x*-axis then the image in the *y*-axis.

    Does the order of reflections make a difference to the outcome?

3   a   Find a single transformation that
        will map:

        i   B to C   ii   D to C
        iii A to B   iv   B to D
        v   A to C   vi   C to A
        vii D to B

    b   Find a combination of two
        transformations that will
        map:

        i   A to D   ii   B to D
        iii A to C   iv   C to D

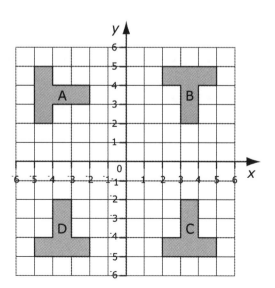

1   Draw a polygon on a 3 by 3 grid.
    Write down:
    ◆   the number of sides
    ◆   whether the polygon is regular
    ◆   the name of the polygon
    ◆   the symmetry properties.

    For example, this shape:

    ◆   Has 4 sides.
    ◆   It is regular.
    ◆   It is a delta, or arrowhead.
    ◆   It has one line of symmetry.

    Repeat for as many different polygons as you can find on a
    3 by 3 grid.

2   Repeat question **1** on a 5 by 5 grid.

    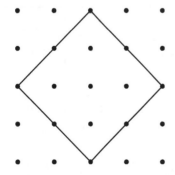

    There are lots of different shapes you can make.
    Try to concentrate on making shapes that have symmetry.

3   Investgate further on a 6 by 6 grid.

Here is a shape on cm squared paper:

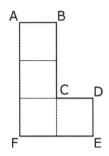

1 Write down:

    a the lengths of each of the edges

    b the area of the shape

    c the lengths of the edges following an enlargement of scale factor 2

    d the area of the shape following an enlargement of scale factor 2.

2 Show how the shape will tessellate to make an enlargement of scale factor 3.

Fit copies of the small shape into the enlarged shape, leaving no gaps or overlaps.

This grid shows a shape:

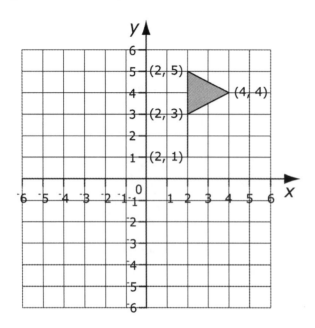

For each of these enlargements you should copy the shape onto a new grid.

Enlarge the shape:

1   by a scale factor of 2 with centre (4, 4)

2   by a scale factor of 2, centre (1, 3)

3   by a scale factor of 2, centre (0, 0)

4   by a scale factor of 2, centre (3, 4).

Are all your shapes the same size?

Explain your answer.

Level 5

This shape is called an **L-triomino.** It is made from three squares.

This shape is made from two L-triominoes. They do not overlap. It has only **one line** of symmetry.

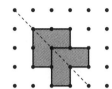

You may use a mirror or tracing paper to help you in this question.

a On a square grid draw a **different** shape made from two L-triominoes which do not overlap. It must have only **one line** of symmetry. *1 mark*

b On your grid draw a shape made from two L-triominoes which do not overlap. It must have **two lines** of symmetry. *1 mark*

This shape is made from two L-triominoes which do not overlap. It has **rotational** symmetry of order **two**.

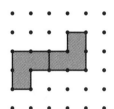

c On a square grid draw a **different** shape made from two L-triominoes which do not overlap. It must have **rotational** symmetry of order **two**. *1 mark*

d On a square grid draw a shape made from two L-triominoes which do not overlap. It must have **two** lines of symmetry **and rotational** symmetry of order **two**. *1 mark*

Level 6

Julie has written a computer program to transform pictures of tiles.

There are **only two instructions** in her program,

Reflect vertical

or

Rotate 90° clockwise.

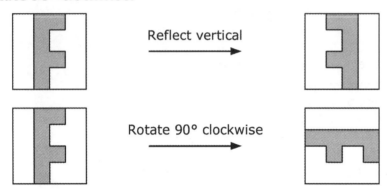

a Julie wants to transform the first pattern to the second pattern.

First pattern   **Second pattern**

Copy and complete the instructions to transform the tiles B1 and B2.
You must use only **Reflect vertical** or **Rotate 90° clockwise**.

A1 *Tile is in the correct position.*

A2 *Reflect vertical, and then Rotate 90° clockwise.*

B1 *Rotate 90° clockwise, and then.....................................*

B2 *...*

2 marks

continued

b Paul starts with the first pattern that was on the screen.

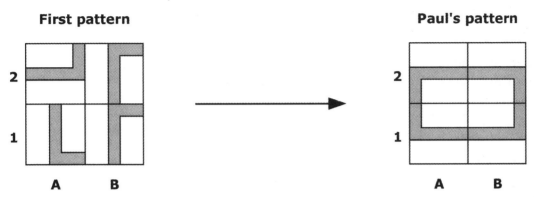

First pattern

Paul's pattern

Copy and complete the instructions for the transformations of A2, B1 and B2 to make Paul's pattern.

You must use only **Reflect vertical** or **Rotate 90° clockwise**.

A1 *Reflect vertical, and then Rotate 90° clockwise.*

A2 *Rotate 90° clockwise, and then* ....

B1 ....

B2 ....

Remember:
◆ In Hot Crosses, the sums of the expressions in the horizontal and vertical lines are the same.

1 Construct two equations for each Hot Cross.
Choose which one to solve first, and work out the value for each letter.

a

	$3x$	
$3y$	7	$2y$
	1	

line total = 17

b

	$2q$	
$5p$	3	$2p$
	q	

line total = 24

c

	15	
$2m$	$3m$	7
	$3n$	

line total = 42

d

	$6r$	
4	$2r$	$3t$
	4	

line total = 20

2 In this magic square, each horizontal, vertical and diagonal line adds up to the same total.

$2x$	$2y - z$	$3y + 5z$
$2y + x$	$2y + z$	$x - 2z$
	$2y + 3z$	$4x$

$x = 3$
$y = 5$
$z = {}^-1$

a Work out the line total.

b Write an expression for the empty square.

1 Copy and complete these addition grids.

a

+	2x + 1	3x
5x + 4		
7x		

b

+	3x – 2	5 – 2x
3x + 1		
x + 7		

c

+		7x – 5
2x	3x + 2	
		10x

d

+		
	8x – 4	5x + 3
	6x	3x + 7

2 In these questions, use the information to make an equation.
Solve this equation to find the value of x.

a

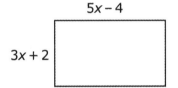

5x – 4

3x + 2

Perimeter = 76 cm

b

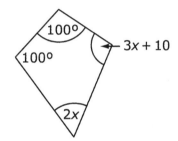

100°

100°

3x + 10

2x

c Three numbers have a mean of 18.
Two of the numbers are 12 and 15.
Find the value of the missing number x.

3 In this arithmagon, the value in a
rectangle is the sum of the
expressions in the circles either side.
You have enough information to find
the value of x.

Find the value of x and the missing
totals on the other two rectangles.

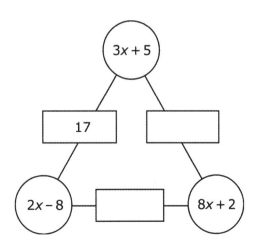

3x + 5

17

2x – 8

8x + 2

> **Remember:**
> ♦ When solving equations you must always keep both sides of
> the equation equal or in balance.

1 Use the balance method to solve each of these equations:

a $5r - 2 = 3r + 6$ **b** $6s + 5 = 3s + 26$

c $2t + 9 = 15 - t$ **d** $4u + 7 = 2u + 18$

e $5v - 7 = 3v - 15$ **f** $16 - 3t = 7 - t$

2 In each Hot Cross the horizontal total is equal to the vertical
total. Form equations for each of these and solve them:

a

	$4x$	
$3x$	7	$2x$
	5	

b

	$9x$	
28	$3x$	$2x$
	7	

c

	$12x$	
$7x$	$8x$	31
	6	

3 In these questions use the information to make an equation.
Solve the equation to find the value of x.

a

$3x - 8$

$x + 15$ square

b

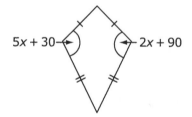

$5x + 30$ $2x + 90$

c

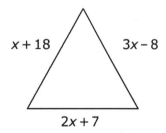

$x + 18$ $3x - 8$

$2x + 7$

Perimeter $= 4x + 35$

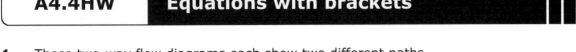

1   These two-way flow diagrams each show two different paths
that lead to the same result.
Form equations from these two-way flow diagrams and then
solve to find the value of x.

a

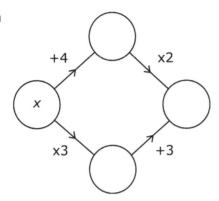

b

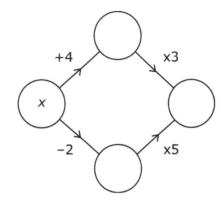

c

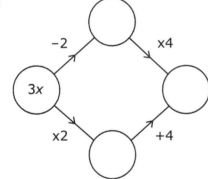

2   Solve these equations (multiply out the brackets first).

   a $5(2x + 3) = 4(3x + 2)$

   b $3(4x - 1) = 2(3x + 3)$

   c $6(3x + 5) = 4(5x + 2)$

3   Here are three expressions:

| $3(2x + 3)$ | | $5(x + 4)$ | | $4(x + 8)$ |

Two of them are equal

Make three different equations, each using only two of the
expressions.

Solve your equations to find the three possible values of x.

Here is a sequence of patterns:

1 Draw the next pattern in the sequence.

2 Copy and complete this table:

Pattern number	1	2	3	4	5	6	7
Number of dots	1			19			

3 Explain in words how the number dots is increasing in this pattern.

4 Here are three formulae, where *n* is the pattern number:

$n^2 + 2n - 2$

$n^2 + n - 1$

$n^2 + 3n - 3$

Which formula represents this number sequence?

5 Use this formula to find out how many dots are in the 10th and 15th patterns.

6 What is the pattern number for the smallest pattern that uses over 500 dots?

1 Work out the value of each expression if:

$a = 4, b = 3, c = 7$

a $a - bc$

b ab^2

c $ab + c$

d $b + c^2$

2 Add a pair of brackets to each expression in question 1, to change the value.
Work out the new value for each.

3 **Match the expression**

For each of these statements write an expression/equation for the information given and match it with one given.

In 5 hours three cyclists Mo, Bash and Deep travel these distances:

Mo p km

Bash q km

Deep r km

Write expressions or equations for this information:

a Deep cycles 30 km more than Bash.

b Mo and Bash cycles 320 km between them.

c Deep cycles 10 km further than Mo.

d The total distance travelled by all three is 500 km.

e Mo cycles 20 km further than Bash.

Use your expressions and equations to find the values of p, q and r.

Level 5

A teacher has a large pile of cards.

An expression for the **total** number of cards is **6n + 8**.

a The teacher puts the cards in two piles.

The number of cards in the first pile is **2n + 3**.

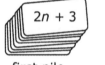

first pile   second pile

Write an expression to show the number of cards in the second pile.

1 mark

b The teacher puts all the cards together.

Then he uses them to make **two equal piles**.

Write an expression to show the number of cards in one of the piles.

1 mark

c The teacher puts all the cards together again,
then he uses them to make two piles.

There are **23** cards in the first pile.

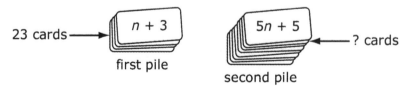

How many cards are in the second pile?
Show your working.   *2 marks*

Class 9H were playing a number game.

Elin said:

> Multiplying my number by 4 and then subtracting 5 gives the same answer as multiplying by 2 and then adding 1.

a   Lena called Elin's number x and formed an equation:

$$4x - 5 = 2x + 1$$

Solve this equation and write down the **value** of x.
Show your working.

2 marks

Aled said:

> Multiplying my number by 2 and then adding 5 gives the same answer as subtracting my number from 23.

b   Call Aled's number y and form an equation.

1 mark

Work out the **value** of Aled's number.

1 mark

Hannah has been planning a project about transport.

She noticed that lots of students are driven to school by a parent.
Often, there are just two people in the car.

Hannah thinks that the school should encourage people to share lifts,
or find other ways of getting to school.

As a first step, she wants to find out how big the problem is.
She writes down a hypothesis:

> Too many people arrive at school by car.
>
> This causes unnecessary congestion and pollution.

Hannah's teacher says:

> What does 'too many' mean? You will need to be specific and include some numbers!

Hannah's friend says:

> The second part of your hypothesis may be true, but do you really need to include it?

1 Rewrite Hannah's hypothesis. Try to improve it by considering the suggestions made by Hannah's teacher and by her friend.

2 Explain how you would test your hypothesis. You should describe:

◆ The type of data you would use.
You should say whether the data you use would be secondary or primary.

◆ How and where you would collect your data.

◆ How you would analyse and display your data.

1   Draw a pie chart to represent the data in this table.

Hair colour	Brown	Fair	Red	Black
Number of people	8	6	1	5

2   Jasna and Sophie are doing a project about birds.
They are interested in the connection between the lengths
of various species of birds, and their weights.

They collect this data:

Species	Weight (grams)	Length (cm)
Greylag goose	3500	80
Mallard	1050	58
Kestrel	210	33
Oystercatcher	500	43
Sanderling	60	20
Herring gull	975	55
Woodpigeon	475	40
Wren	10	10
Mistle thrush	125	26
Cormorant	2400	90
Turnstone	110	22
Blackbird	90	25

Draw a scatter diagram to show the data.

Write a paragraph to explain the general pattern shown
in the diagram, and any unusual features.

1   The scatter diagram shows the engine size and fuel consumption
    (miles per gallon) for a selection of cars.

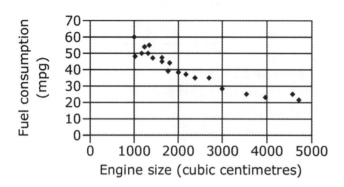

Write a paragraph to describe the relationship between engine
size and fuel consumption shown in the scatter diagram.

2   These pie charts show the types of houses on two different streets.

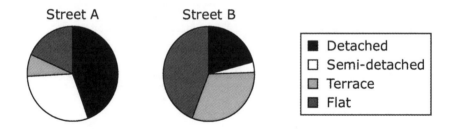

Decide whether each of these statements is true, false, or
impossible to decide. Explain your reasoning.

a   Street B has more terraced houses than Street A.

b   More than a quarter of the homes on Street A are
    semi-detached houses.

c   Detached houses were the most common type of home
    on Street B.

d   If you count all the homes in both streets, there were
    more flats than terraced houses.

1   The table shows the speeds (in miles per hour) of 30 cars
travelling down a stretch of road.

35	57	56	45	52	39	49	47	52	46
49	40	47	44	45	42	52	45	48	59
47	47	39	50	51	36	49	49	40	51

Copy and complete this stem-and-leaf diagram to show this data.

50	
40	
30	

2   Use the diagram you drew for question **1** to find:

a   the median speed of the cars

b   the range of the speeds.

3   A team of biologists measure the length of 25 fish.
The table shows the data they collect (in cm, to 1 dp).

8.8	7.6	8.6	9.8	9.2
7.3	6.8	9.4	5.5	9.4
7.9	9.6	7.3	9.2	8.4
7.5	7.2	11.2	11.2	7.5
8.1	8.0	8.8	5.5	11.1

a   Draw a stem-and-leaf diagram for the data.

b   Use your diagram to find the median length.

c   Use your diagram to work out the range of the lengths.

How big is your mouse?

A team of scientists is investigating colonies of mice living in two different locations. They think that the mice in the kitchen tend to be bigger than the mice in the cellar.

To test this hypothesis, they capture a sample of mice from each location, and measure their body lengths.

The data are shown in the table; the lengths are given in centimetres, to 1 dp.

Kitchen

7.4	5.8	6.7	8.7	6.0
8.0	6.2	6.8	5.3	9.9
5.7	7.9	9.7	6.8	10.9
8.6	7.7	8.5	10.1	8.2
5.1	8.4	5.4	9.1	6.3

Cellar

9.3	7.7	6.8	9.1	6.5
8.6	7.7	4.2	4.1	7.0
7.6	9.3	6.4	4.8	7.7
8.4	6.0	8.7	8.5	4.9
8.5	4.4	5.0	3.8	6.3

1 Draw a back-to-back stem-and-leaf diagram to show both sets of data.

2 Use your diagram to find the median length of the mice at each location.

3 Use your diagram to work out the range of the lengths at each location.

4 Calculate the mean body length of the mice at each location.

5 Explain whether you think the scientists' hypothesis (that the mice in the kitchen are bigger than those in the cellar) is justified by the diagram you drew and the statistics you calculated.

Sara's bloomin' garden

Sara is a keen gardener, and she is particularly proud of her lawn.

She wants to know whether a new type of lawn fertiliser will make her lawn grow better.

Sara decides to carry out an experiment:
◆ She marks out two equal patches of lawn.
◆ She treats one of the patches with the fertiliser, and leaves the other one untreated.

Sara says:

> I'm going to mow both patches every week for 12 weeks, and I'm going to weigh the grass clippings from each patch. If the fertiliser is working, the treated patch should produce more grass!

The table shows the results of Sara's experiment.

Weight of grass clippings – in kg, to 1 dp

Week	1	2	3	4	5	6	7	8	9	10	11	12
Treated	5.5	6.5	6.7	8.3	7.3	8.5	7.3	10.5	10.4	5.2	8.0	9.6
Untreated	4.3	5.9	6.2	4.4	4.2	5.4	2.5	8.2	5.4	3.2	6.2	6.0

Sara writes a weekly gardening column for her local newspaper.

She decides to write about her experiment.

Write a report that Sara could use in her newspaper column.

You should include appropriate statistics and diagrams, and you should state your hypothesis and conclusions clearly.

Level 5

This advert was in a newspaper:

93% of us
drop litter
every day.

Gang up on litter

Do your bit. Use a bin.

It does not say how the advertisers know that 93% of people drop litter every day.

Some pupils think the percentage of people who drop litter every day is much lower than 93%.

They decide to do a survey.

a Jack says:

> We can ask 10 people if they drop litter every day.

Give two **different** reasons why Jack's method might not give very good data. *2 marks*

b Lisa says:

> We can go into town on Saturday morning.
>
> We can stand outside a shop and record how many people walk past and how many of those drop litter.

Give two **different** reasons why Lisa's method might not give very good data. *2 marks*

Three types of mice might come into our homes.

Some mice are more likely to be found in homes far from woodland.
Others are more likely to be found in homes close to woodland.

The bar charts show the **percentages of mice** that are of each type.

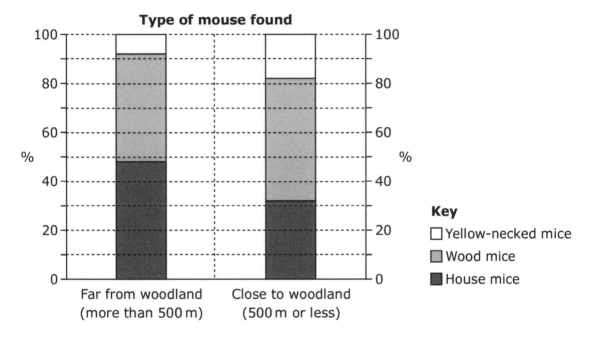

Use the bar charts to answer these questions.

a About what percentage of mice in homes **close to woodland**
  are **wood mice**? *1 mark*

b About what percentage of mice in homes **far from woodland**
  are **not** wood mice? *1 mark*

c The **black** bars show the percentages for house mice.
  One of the black bars is taller than the other.

  Does that mean there **must be more** house mice in homes
  far from woodland than in homes close to woodland?

  Explain your answer. *1 mark*

1 Calculate these, using a mental or written method.

    a   15.14 + 6.1 + 0.07   **b**   13.7 − 2.93 + ⁻1.4

    c   33.3 - ⁻253.8 + 87.65   **d**   7.26 + 3.02 + ⁻4 + 0.09

    e   419 − 3.003 − 2.3   **f**   0.0848 + 137.106 + 2593

    g   432.11 + 0.051 − 17.69   **h**   3.048 km - 0.079 km

    i   9006.9 m - ⁻42.8 m − 0.7 m   **j**   176.7 litres + 0.58 litres + 19 litres

2 Copy and complete these number walls.

    a

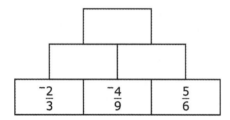

    b

3   **Puzzle**

    a   Afzal, Brian, and Jeff buy three different salads to share.
        The salads are identical in weight.

        Afzal has $\frac{1}{2}$ of the first salad, $\frac{1}{5}$ of the second salad and $\frac{7}{20}$ of the third salad;

        Brian has $\frac{1}{6}$ of the first salad, $\frac{1}{3}$ of the second salad and $\frac{2}{5}$ of the third salad;

        Jeff has $\frac{1}{3}$ of the first salad, $\frac{7}{15}$ of the second salad, and $\frac{1}{4}$ of the third salad.

        Who had the most salad in total?

        Who had the least salad?

        Explain and justify your answers.

    b   Invent a similar word problem of your own.

1 You may need to make some jottings to calculate these:

a ⁻25 x 0.7

b ⁻24 ÷ 0.6

c $\frac{4}{7}$ x ⁻28

d ⁻8 ÷ ⁻0.04

e $\frac{3}{8}$ x 36 cm

f ⁻4.8 ÷ 0.3

g ⁻12 x 4.3

h 3.6 x 1.25

2 **Puzzle**

Use each of the digits from 0 to 9 once only to complete these three calculations:

a

$$\Box\Box \div 0.\Box =$$

b

$$\frac{\Box}{\Box} \times \Box\Box =$$

c

$$\Box\Box \times 0.\Box =$$

d Add up the total of the three calculations.

By making up the calculations with different digits:

e Make the largest total you can.

f Make a total as near to 200 as you can.

g Make a total of exactly 462.8.

93

1 Solve each of these calculations.

You will need to decide whether to use a mental, written or calculator method.

Where appropriate, give your answer to 2 dp.

> **Remember:**
>
> The order of operations is:
>
> **Brackets** first
>
> **Powers** or **indices** next
>
> **Multiplication** and **division**
>
> **Addition** and **subtraction**

a $(25 - 9) \times (12 + 2)$

b $\dfrac{(1.5 \times 4)^2}{9}$

c $\dfrac{5^2 - 10}{2.4 + 2.6}$

d $6 + 4(8 - 3)$

e $24 \div 4 \times 3$

f $6^2 \div [8 - (4 - 2)]$

g $\sqrt{(3.8^2 \pm 1.2^2)}$

h $\dfrac{(3.2 - 4.2)^2}{\sqrt{(96.4 + 0.4 \times 9)}} + 2.7 \times 0.5$

2 **Puzzle**

Insert the correct operations to make these calculations correct:

a $3 \; \Box \; 4 \; \Box \; 6 \; \Box \; 2 = 15$

b $40 \; \Box \; 3 \; \Box \; 8 = 16$

c $5[3 \; \Box \; (4 \; \Box \; 7)] = 0$

d $\sqrt{[2^3 \; \Box \; 5 \; \Box \; (2 \; \Box \; 1.7)]} = 5$

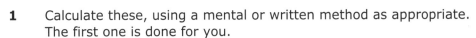

1 Calculate these, using a mental or written method as appropriate.
The first one is done for you.

a 0.3 x 0.6

a 0.3 x 0.6

0.3 x 0.6 = 3 x 6 x 0.1 x 0.1

= 18 x 0.01

= 0.18

b 2.67 x 9

c 8.52 x 6

d 12.4 x 0.3

e 27.3 x 2.5

f 16.8 x 2.9

g 34.9 x 7.4

h 18.7 x 6.1

i 11 x 2.77 m

j 34.6 km x 8.3 km

2 **a** 'Butcher's block' worktops cost £88.40 per square metre.
How much would it cost to cover this workspace?

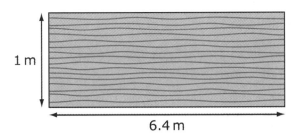

1 m

6.4 m

b Haricot beans cost £1.40 per kg.

How much would 32.6 kg of haricot beans cost?

c The ratio of the height of Paul to the height of Shoaib is 6 : 5.
If Shoaib is 1.6 m, how tall is Paul?

d Sisal carpets cost £23.70 per square metre.

How much would it cost to carpet an area of 9.8 m² with sisal?

3 **Puzzle**

The numbers 4.3, 7.5, 3.4, 2.6, 4.2 and 7.9 are printed on six cards.
Use the numbers to solve each of the following problems.
Each number can be used only once in each problem.

a ? x ? + ? = 28.1

b ? x ? + ? x ? = 77.31

c ? x (? + ?) + ? x ? = 78.18

1 **a** Copy and complete Damien Allbran's shopping bill:

Cost per item	Number of items	Total cost
£3.25		£32.50
£0.70		£28.00
£1.40		£63.00
	6	£14.16
£4.26	5	£21.30
£2.60		£39.00
	14	£19.32
	Total	

Remember to fill in the total.

b How many items did Damien buy?

2 Calculate these (giving your answer to 1 dp where appropriate):

a 130 ÷ 2.7 **b** 115 ÷ 4.6

c 49.6 ÷ 18 **d** 21 ÷ 2.8

e 34.7 km ÷ 8 **f** 328 ÷ 1.8

g 440 m ÷ 6.4 m **h** 36.5 days ÷ 1.5 hours

3 Calculate these (giving your answer to 1 dp where appropriate):

a 15.5 kg of bananas cost £27.90.

What is the price of 1 kg of bananas?

b Linton spends £10.23 on 4.65 m of tinsel.

How much does 1 m of tinsel cost?

c Emma is doing a sponsored silence.

She will earn £3.70 for each hour she remains silent.

If she earns a total of £45.88, for how many hours and minutes was she silent?

1   **Investigation**

In the UK many quantities used to be measured in imperial units.
Here is part of a conversion table for changing inches (imperial units)
into centimetres (metric units).

Inches	Centimetres
2	5.08
5	12.7
10	25.4
25	63.5

a   What is the scale factor used to change 2 inches into 5.08 centimetres?

b   What is the scale factor used to change 10 inches into 25.4 centimetres?

c   Write down what you notice. Can you explain what you have noticed?

d   Use your scale factor to change 17 inches into centimetres.

e   How would you change 20 cm into inches?

2   Investigate scale factors between other sets of metric and imperial units.
This table will help you.

Metric		Imperial
1 metre	~	3 feet
8 km	~	5 miles
1 kg	~	2.2 lbs
30 g	~	1 ounce
1 litre is just less than 2 pints		

a   Design a poster to illustrate your findings.

b   Convert each of these measurements into imperial units:

    i   750 ml of olive oil

    ii   500 g of sugar

    iii   40 litres of petrol

    iv   the 1500 metres race (in feet and miles)

    v   a 30 cm ruler. (Remember: 2 inches ~ 5.08 cm.)

Level 5

Bill, Ravi, and Eric are three divers in a competition.

Each type of dive has a **dive rating**.

Easy dives have a **low** rating; **hard** dives have a **high** rating.

Every dive is marked by five judges who each give a **mark out of 10**.

How to calculate the score for a dive:

1 Look at all five marks. Remove the highest and the lowest marks.

2 Add together the middle three marks to give a total.

3 Multiply this total by the dive rating.

---

a Bill does a dive with a dive rating of 3.34.
The judges give the marks   7.0   7.5   8.0   8.0   8.5
What is Bill's score? *1 mark*

b Ravi scored 82.68 on his first dive.
The dive had a dive rating of 3.18.
What was the **total** of the middle three marks given by the judges? *1 mark*

c Eric is getting ready to take his final dive.
He needs to score at least 102.69 to win the competition.
Eric decides to do a dive with a dive rating of 3.26.

Explain why Eric has made a poor decision.
Show your working. *2 marks*

Level 6

The table shows information about some countries in Europe.

Country	Area in km^2	Population
France	543 965	57 456 000
Netherlands	41 785	15 129 000
Spain	504 782	39 952 000
United Kingdom	244 103	57 561 000

a Which country has the **most** people for each km^2?
Which country has the **fewest** people for each km^2?

Copy this table, and write down these countries and their values for the 'number of people for each km^2'.

	Country	Number of people for each km^2
Most people for each km^2		
Fewest people for each km^2		

2 marks

b Look at the information for the United Kingdom.
Imagine that the area of land was shared out equally amongst all the people.

Calculate how much land, in m^2, each person would get.

1 km^2 = 1 000 000m^2. *2 marks*

> **Remember:**
> ◆ To expand brackets you multiply each term inside the bracket by the term outside.

1 Simplify these expressions:

  a $5x + 7 - 2x + 3$ **b** $8x - 7 + 2x - 5$

  c $3x + 2(x - 4)$ **d** $5(2x - 8) - 31$

  e $6x - 7(2x + 3)$ **f** $9x + 4(3x - 7)$

2 Here are seven expression cards, with five turned face down.

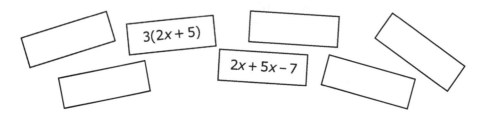

$3(2x + 5)$

$2x + 5x - 7$

  a Three of the expression cards are equivalent.

  Choose one of the expressions given, and write two expression cards that are equivalent to it.

  b For the other expression given, write one equivalent expression card.

  c The other two expression cards have expressions that are half the value of the two expressions given.

  Write these two expression cards.

3 **a** Write an expression for the perimeter of this triangle.

  b Write and expression for the area of this triangle.
  Expand the bracket in your expression.

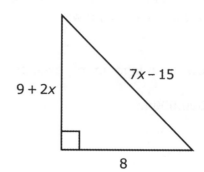

$7x - 15$

$9 + 2x$

8

> **Remember:**
> ♦ To solve an equation you must always keep the balance.

1 Copy and complete this spider map, by following the instructions on each leg.

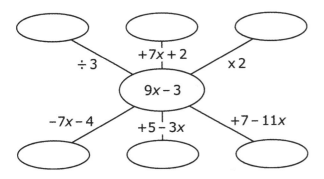

2 Each of these diagrams show two expressions that balance.
Write each balance as an equation and solve it.

a $3n - 4$ $n + 10$ **b** $5p - 3$ $2p + 12$

c $5t + 6$ $24 - t$ **d** $2(3p + 7)$ $10p - 6$

3 Here are three expressions:

A: $3x - 7$ B: $5x + 3$ C: $4x - 20$

a Find x if A is equivalent to B.

b Find the value of x if B is equivalent to C.

c Find the value of x if 2A is equivalent to C.

4 Make up your own problem similar to question 3.

1 In an arithmagon, each expression in a square is the sum of the expressions in the circles on either side of it.

a Copy and complete this arithmagon:

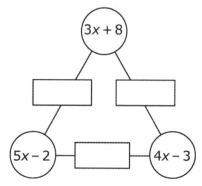

b The top left box is equal to 30.
Use this information to find the value of x.

2 In these two arithmagons, use the information given to find the value of x.

a

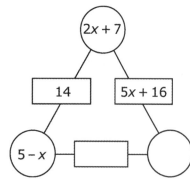

b

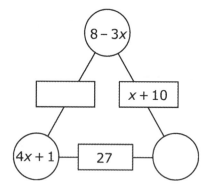

3 Solve these equations:

a $6x - 7 = 26$

b $\dfrac{x}{7} - 3 = 4$

c $2x - 12 = 18 - 2x$

d $3(2x - 9) = 12$

e $\dfrac{2x+7}{5} = 8$

f $3(2x + 3) = 4(x - 2) + 20$

Each of your answers to **a**–**f** should match one of these values:

$x = 16{\cdot}5$ $x = 5{\cdot}5$ $x = 6{\cdot}5$ $x = 49$ $x = 2{\cdot}5$ $x = 1{\cdot}5$ $x = 7{\cdot}5$

One of the values is not used.
Make an equation (using brackets) where x is this value.

Remember:
♦ You can undo division by multiplying.

In each question, one equation has a different solution to the other two.
Find the 'odd' solution for each question.
Use these letters to spell a word.

1 $4r + 6 = 34$, $2s - 9 = 7$, $5t + 3 = 38$

2 $\dfrac{3u - 4}{4} = 5$, $\dfrac{6v + 2}{5} = 4$, $\dfrac{11 + 3w}{4} = 5$

3 $5(a + 7) = 45$, $3(5b - 3) = 21$, $6(4 + 3c) = 42$

4 $2a - 3 = a + 4$, $5b - 12 = 2b + 9$, $4c + 5 = 6c - 3$

5 $\dfrac{4}{d+1} = 2$, $\dfrac{8}{e-2} = 4$, $\dfrac{6}{2f+1} = 2$

6 $\dfrac{5}{r+3} = \dfrac{4}{r+1}$, $\dfrac{3}{s-2} = \dfrac{5}{2s-5}$, $\dfrac{3}{2t+4} = \dfrac{5}{4t+2}$

7 $5r - 6 = 14$, $3(s + 5) = 5s + 3$, $2(3t + 2) = 7(2t - 4)$

Write down the word made by these letters.

8 Add up the **value** of all the odd answers from 1–7.

   a   What is this total?

   b   If $\dfrac{5x + 8}{3}$ = your total from **a**, what is the value of x?

1   **Mix and Match**

Green paint is made by mixing blue and yellow paint.
It is sold in 12 litre tins.

Ratios
Blue : Yellow
2 : 1
1 : 2
5 : 1

dark green   mid green   light green

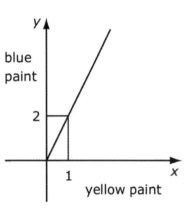

12 litres   12 litres   12 litres
green mix   green mix   green mix

10 blue   8 blue   4 blue
2 yellow   4 yellow   8 yellow

Equations
$y = 5x$
$y = \dfrac{1}{2}x$
$y = 2x$

Match each tin of green paint with its graph, equation and ratio.

a

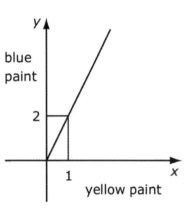

b

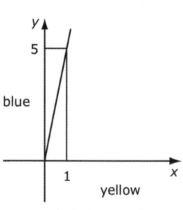

c

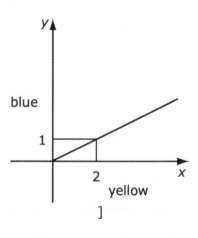

2   Grey paint is made by mixing black and
white paint in the ratio 1 : 3.

On a grid similar to this, sketch a line to
show the mixture of paint.

Write down the equation of the graph.

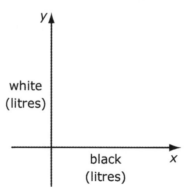

> **Remember:**
>
> ◆ The equation a straight line is of the form $y = mx + c$.
>
> ◆ Gradient $= \dfrac{\text{vertical distance between the points}}{\text{horizontal distance between the points}} = \dfrac{\updownarrow}{\leftrightarrow} = m$
>
> ◆ y-intercept = where the line crosses the y-axis = c

1 For each of these graphs find the gradient (m) and the
y-intercept (c)

a

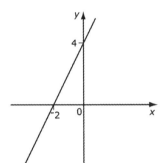

b

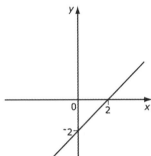

c

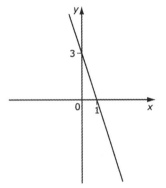

2 For each of the graphs in question 1, write out the equation of
the line in the form $y = mx + c$.

3 Sketch these three graphs on a pair of axes with x from ⁻3 to 3
and y from ⁻5 to 10.
You may find a table of values useful.

a $y = 2x - 1$

b $y = \dfrac{1}{2}x + 3$

c $y = 5 - 2x$

1 This distance-time graph shows Martina's journey on the London Underground from Euston Square to Leicester Square.

    a How long did Martina wait at Euston Square?

    b Where was she at 08.40?

    c Which two stations are the greatest distance apart?

    d What was the total journey time from when the train left Euston Square until it arrived at Leicester Square?

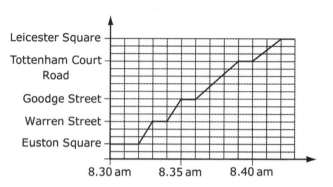

2 Boiling water is placed into three metal cylinders.

    ◆ Cylinder A is left on the bench top.

    ◆ Cylinder B is placed in a bucket of ice.

    ◆ Cylinder C is wrapped in insulation.

The temperature changes are recorded over an hour.
Copy these axes three times and sketch a temperature-time graph for each cylinder.

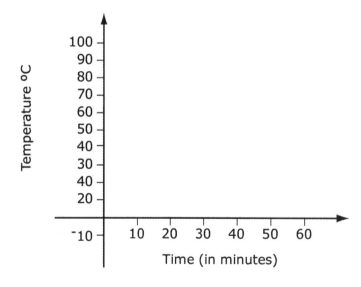

In this exercise all the lengths given are in cm.

1   This diagram shows a square.

    a   By forming an equation, find the value of x.

    b   Find the area of the square.

    c   A rectangle with the same perimeter as this
        square has an area of 96 cm^2.
        Find the length and width of this rectangle.

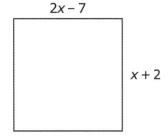

2   In this right-angled triangle, the height
    is twice the base ($3x - 1$)

    a   What is the value of x?

    b   What is the area of the triangle?

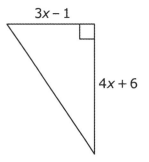

3   In this isosceles triangle the perimeter is 48 cm.

    Find the value for x and the length of the
    third side.

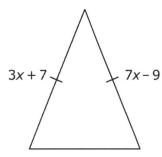

Steve is making a series of patterns with black and grey square tiles.

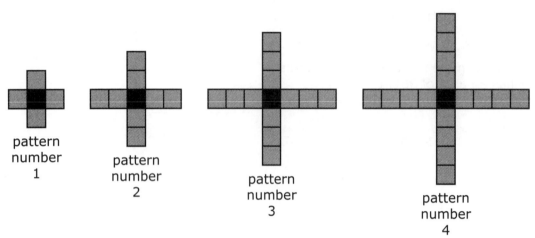

pattern
number
1

pattern
number
2

pattern
number
3

pattern
number
4

a   Each pattern has **1 black tile** at the **centre.**

Each **new** pattern has **more grey tiles** than the one before.

How many **more** grey tiles does Steve add each time he
makes a new pattern?   *1 mark*

b   Steve says: *'The rule for finding the number of tiles in pattern
number N is **number of tiles = 4 × N + 1.***'

The **1** in Steve's rule represents the **black tile.**

What does the **4 x N** represent?   *1 mark*

c   Steve wants to make **pattern number 15.**

How many **black** tiles and how many **grey** tiles does
he need?   *1 mark*

d   Steve uses **41 tiles** altogether to make a pattern.

What is the number of the pattern he makes?   *1 mark*

e   Steve has **12 black** tiles and **80 grey** tiles.

What is the number of the **biggest** pattern Steve
can make?   *1 mark*

Level 6

These straight line graphs all pass through the point (10, 10).

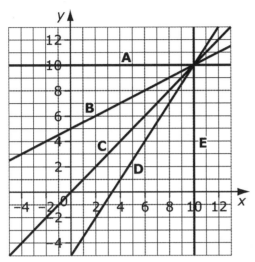

a Copy these sentences and fill in the gaps to show which line has which equation.

line ………….. has equation $x = 10$

line ………….. has equation $y = 10$

line ………….. has equation $y = x$

line ………….. has equation $y = \dfrac{3}{2}x - 5$

line ………….. has equation $y = \dfrac{1}{2}x + 5$ *2 marks*

b Does the line that has the equation $y = 2x - 5$ pass through the point (10, 10)?

Explain how you know. *1 mark*

c I want a line with equation $y = mx + 9$ to pass through the point (10, 10).

What is the value of m? *1 mark*

Here is another path design produced by the Primrose Path Company.

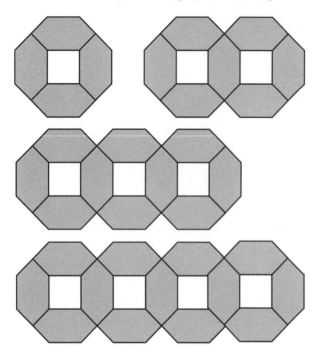

These are the first four patterns in the Squares and Hexagons design.

1 Complete this table to show the number of squares and the number of hexagons needed for the first 10 patterns in this design.

Pattern no.	1	2	3	4	5	6	7	8	9	10
No. of squares	1									
No. of hexagons	4									

2 How many hexagons would be needed for a pattern with 100 squares?
Explain how you worked this out.

3 How many squares would be needed for a pattern with 67 hexagons?
Explain how you worked this out.

4 The company thinks there is something wrong with an order they receive from a customer.

The customer asks for 125 hexagons. Explain why the company thinks the order is wrong.

Here are three different matchstick sequences.

These are the first two patterns in Sequence 1.

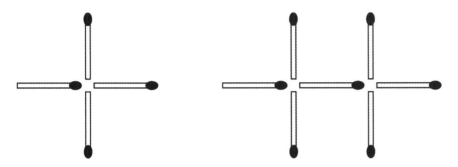

These are the first three patterns in Sequence 2.

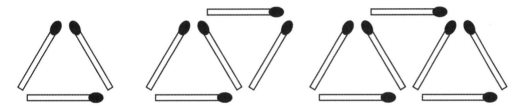

These are the first three patterns in Sequence 3.

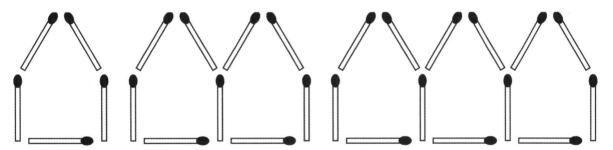

These are the first three patterns in Sequence 3.

For each of these sequences:

a   Make a table to show the number of matches needed for the first five patterns.

b   Show the results from the table as a line graph.

Use the pattern number as the independent variable (*x*), and the number of matches as the dependent variable (*y*).

1 Find a rule, using algebra, for each of these tables.
Use *x* as the independent variable.

a

x	s
1	1
2	4
3	7
4	10
5	13

b

x	s
1	2
2	6
3	10
4	14
5	18

c

x	t
1	21
2	23
3	25
4	27
5	29

Hint:
Your rule will start
s = or *t* =

2 Find a rule, using algebra, for this sequence of patterns:

Use the number of squares as the independent variable, and the number of hexagons as the dependent variable.

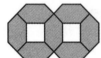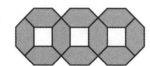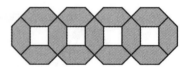

These are the first four patterns in the Squares and Hexagons design.

3 Write down a rule, using algebra, for each of these matchstick sequences.
Use the pattern number as the independent variable, and
the number of matches as the dependent variable.

a

b

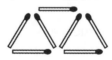

c

> **Remember:**
> To find a unitary ratio, you divide the first number by the second.
> For example:
> The unitary ratio of 23 to 4 is 23 ÷ 4 = 5.75.

1 This table shows values of the variables p and q.

p	1	2	3	4	5	6	7	8	9	10	11
q	14	17	20	23	26	29	32	35	38	41	44
$q:p$											

    a Copy the table, and fill in the final row to show the unitary ratio of q to p

    b Plot a graph to show how the ratio $q:p$ changes as p increases.
Use p as the independent variable, and the unitary ratio $q:p$ as the dependent variable.
Join the points with a smooth curve to show the trend.

    c What do you notice about the curve of the graph?

2 Copy this table:

x	1	2	3	4	5	6
y						
$y:x$						

    a Fill in the values of y, using the rule $y = 2x + 5$.

    b Fill in the third row of the table to show the value of the unitary ratio $y:x$.

    c Draw a graph to show how the ratio $y:x$ changes as x increases.

    d What do you notice about the curve of the graph?

1   The pattern shown is made up of lines and circles.

This arrangement is 'three circles across' and 'two circles down'.

Copy and complete this table to show how many lines there are in arrangements up to five circles across and five circles down.

		Circles across				
	Number of lines	1	2	3	4	5
	1					
	2			17		
Circles down	3					
	4					
	5					

Copy and complete the table, to show how many lines there are in arrangements up to five circles across and five circles down.

2   This diagram shows a pattern made of two different sorts of arrow shapes. This pattern is 'three arrows across' and 'three arrows down'.

(A pattern that was 'one across' and 'one down' would just be a single 'right-pointing' arrow.)

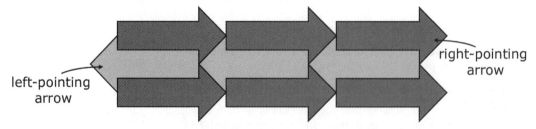

left-pointing arrow

right-pointing arrow

a   Draw a table to show how many 'right-pointing' arrows there are in patterns of any size up to 'six arrows across' and 'six arrows down'.

b   Draw another table for the number of 'left-pointing' arrows in patterns up to 'six across' and 'six down'.

c   Now draw another table to show the **total** number of arrows in the various patterns.

Explain whether the following statements are true. For each one, you should say whether you think the statement is definitely true, definitely false, or whether it is impossible to say.
Explain your reasoning carefully.

1   Brendan collects *Xtreme Skating* magazine.

He has every copy in his collection from number 17 to number 104.

Brendan says: 'I have 87 copies, because 104 – 17 = 87.'

2   Safina is investigating polygons and diagonals from each vertex.

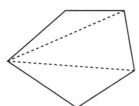 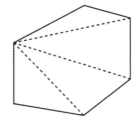

She says:

'A pentagon has 5 sides and 2 diagonals. A hexagon has 6 sides and 3 diagonals. So a polygon with n sides must have n-3 diagonals.'

3   Catriona is collecting temperature data.

She measures the temperature in the school canteen at noon each day.

Day	Mon	Tue	Wed	Thu	Fri
Temperature (°C)	21	20		18	17

Catriona was away on a school trip on Wednesday, so she did not record the temperature that day.

She says:

'It doesn't really matter about the Wednesday temperature; you can see from the other temperatures that it must have been 19°C on Wednesday.'

Helen has these eight rods.

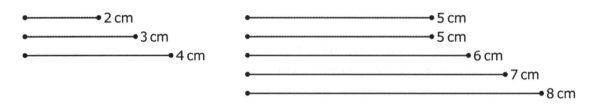

2 cm   3 cm   4 cm   5 cm   5 cm   6 cm   7 cm   8 cm

She can use **5** of her rods to make a **rectangle**.

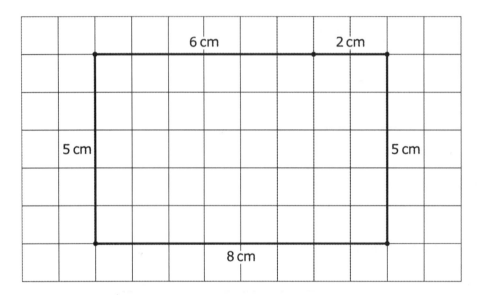

6 cm   2 cm   5 cm   5 cm   8 cm

a   Show on a grid how to make a **different rectangle** with a
different shape with **5** of Helen's rods.   *1 mark*

b   Show how to make a rectangle with **6** of Helen's rods.   *1 mark*

c   Show how to make a **square** with all **8** of Helen's rods.   *1 mark*

Level 6

a   The label on yoghurt A shows this information.

> **Yoghurt A   125 g**
>
> Each 125 g provides
>
> | | |
> |---|---|
> | Energy | 430 kJ |
> | Protein | 4.5 g |
> | Carbohydrate | 11.1 g |
> | Fat | 4.5 g |

How many grams of **protein** does **100 g** of yoghurt provide?

Show your working.

2 marks

b   The label on yoghurt B shows different information.

> **Yoghurt B   150 g**
>
> Each 150 g provides
>
> | | |
> |---|---|
> | Energy | 339 kJ |
> | Protein | 6.6 g |
> | Carbohydrate | 13.1 g |
> | Fat | 0.2 g |

A boy eats the same amount of yoghurt A and yoghurt B.

Which yoghurt provides him with more **carbohydrate**?

Show your working.

2 marks

You can draw cubes on isometric paper like this:

Draw vertical lines vertically:

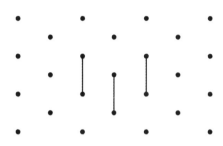

Make sure your paper is this way up!

Draw the top face:

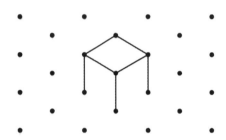

Complete the bottom:

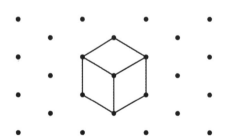

On isometric paper, draw all the different arrangements of:

1 2 cubes

2 3 cubes

3 4 cubes.

For each arrangement, state the number of faces, edges and vertices.

You can make different shapes from five multilink cubes:

For example:

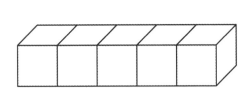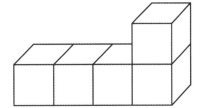

You can draw these shapes on isometric paper:

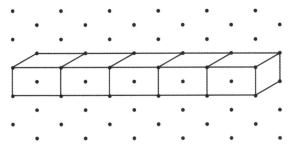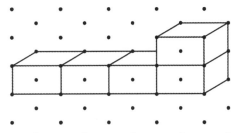

You can draw the views of the shapes on squared paper:

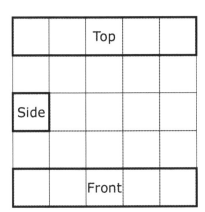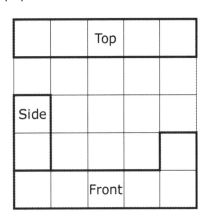

a   On isometric paper, draw as many different arrangements
as you can find of five cubes.

b   For each arrangement, draw:
- ◆   the plan view (from the top)
- ◆   the side elevation and
- ◆   the front elevation.

> **Remember:**
> Volume of a cuboid = length x width x height
> Surface area of a cuboid = 2 x plan area + 2 x side area + 2 x front area

1 Find the volumes and surface areas of these cuboids.

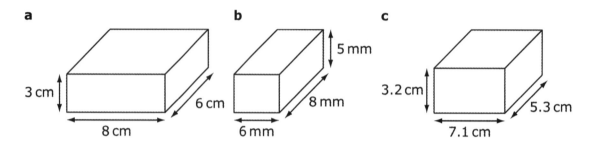

a

3 cm
8 cm
6 cm

b

5 mm
6 mm
8 mm

c

3.2 cm
7.1 cm
5.3 cm

2 Calculate the volume of this prism.

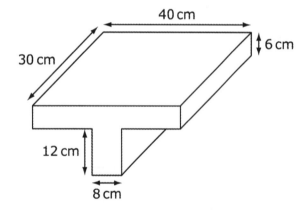

40 cm
6 cm
30 cm
12 cm
8 cm

3 Calculate the volume of this prism.

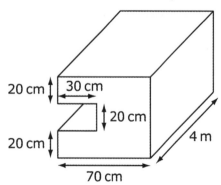

20 cm
30 cm
20 cm
20 cm
4 m
70 cm

1   **a**   Jon has drawn a scale diagram of his bathroom:

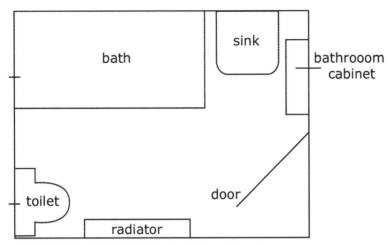

Choose a suitable scale and draw a scale diagram of your bathroom.

Your drawing should include the features of the bathroom, such as bath, toilet, sink, radiator, bath mat and so on.

b   Suggest alternative arrangements of your bathroom.

2   Find the scale of each of these in the form of 1:*n*.
The first one is done for you.

a   A map of a lake using 10 cm to represent 100 m.
The scale is 10:1000 = 1:100

b   A map of a village using 5 cm to represent 1 km.

c   A map of Sheffield using 2 cm to represent 1 km.

d   A map of Europe using 2 cm to represent 200 km.

e   A plan of your school using 2 cm to represent 10 m.

f   A plan of a church with 1 inch representing 1 yard.

g   A plan of a sports ground with 1 inch representing 5 yards.

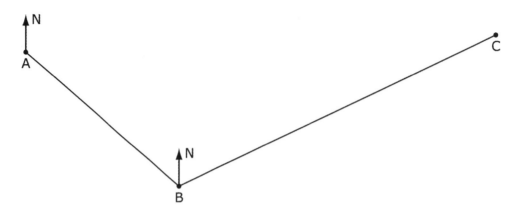

S4.5HW   **Bearings**

1   This diagram is drawn using a scale of 1 cm to represent 2 km.

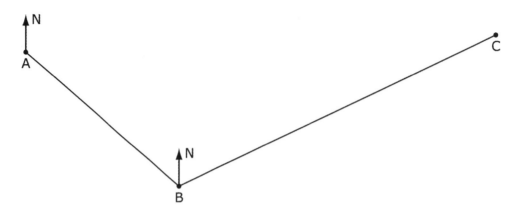

Find:

a   the distance AB

b   the distance BC

c   the distance AC

d   the bearing of B from A

e   the bearing of A from B

f   the bearing of C from B

g   the bearing of B from C

h   the bearing of C from A

i   the bearing of A from C.

> **Remember:**
> Bearings are measured clockwise from North.

2   **a**   Draw a scale diagram using 1 cm to represent 2 km to show this journey:

A speedboat travels on a bearing of 120° for 12 km followed by a bearing of 070° for 10 km.

b   Find the speedboat's distance from its starting point.

c   Find the speedboat's bearing from its starting point.

1 Find the midpoints of each of the lines on this grid.

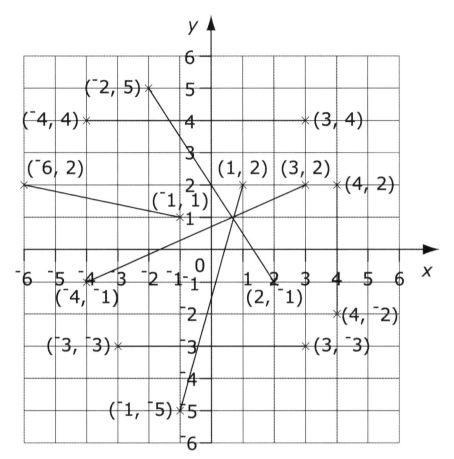

2 Find the midpoints of the lines joining:

a (6, 0) to (10, 0)

b (4, 6) to (4, 8)

c (6, 6) to (12, 12)

d (6, ⁻6) to (⁻2, 0)

e (4, ⁻7) to (⁻7, ⁻1)

f (⁻5, 4) to (3, ⁻2)

g (6, 10) to (20, 30)

h (20, 10) to (10, 20).

1   Describe the journey from your Maths classroom to the Dinner Hall.

Be as clear and precise as you can.

Imagine you are describing the journey to someone who does not know the school.

2   Sketch the path of each of these objects.

a   A bird flying from A to B.

b   An aeroplane taking off.

c   A ball thrown straight up in the air and then caught.

d   A conker on a string.

e   The hour hand on a watch.

3   Two points A and B are exactly 8 cm apart.

On squared paper, show the locus of points that are equidistant from A and B.

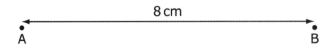

8 cm

A   B

4   Two trees are 14 m apart.

A path is built equidistant from the two trees.

a   Sketch one position of the path.

The path is 2 m wide.

b   Draw a scale drawing to show the position of the path.

1 **a** Make a sketch of the locus of points 5 cm away from a point x.

   b Construct the locus.

2 A radio mast can transmit within a 20 km radius.

   Make a scale drawing to show the region that can receive the
   radio signal.

   Clearly state the scale you are using.

3 Farmer Ted has only one field which is shaped as in this diagram.

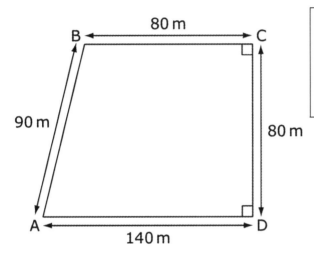

Remember:

The locus of points that are the
same distance from two adjacent
edges is the bisector of the angle
between the edges.

He decides to use fences to keep his animals separate.

He constructs one fence that is equally distant from edges
AB and AD.

He constructs another fence that is equally distant from
edges CB and CA.

Draw a scale diagram to show the different regions his
fences will make.

1 Why is this triangle impossible to construct?

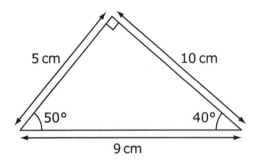

5 cm   10 cm
50°   40°
9 cm

2 Construct this triangle ABC.

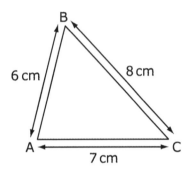

B
6 cm   8 cm
A   7 cm   C

Measure the angles of the triangle.

3 Construct a triangle with sides 12 cm, 8 cm and 6 cm.

4 Try to construct a triangle ABC where AB = 9 cm, AC = 4 cm and BC = 5 cm.

What goes wrong?

5 Construct this quadrilateral.

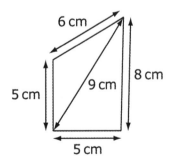

6 cm
8 cm
9 cm
5 cm
5 cm

6 Can you construct a triangle with sides 20 cm, 10 cm and 10 cm?

Give a reason for your answer.

Level 5

This prism is made from 6 cubes.

The piece of paper below fits exactly around the **sides** of the prism.

The **dashed** lines are **fold** lines.

a A different prism is made from 10 cubes.

Copy and complete the diagram below to show a piece of paper that fits exactly around the sides of the 10 cube prism.

Show all folds as dashed lines.

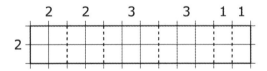

2 marks

b The piece of paper below fits exactly around the sides of a 14 cube prism.

Draw this 14 cube prism on an isometric grid, like this one:

3 marks

Level 6

The diagram shows a model made with **nine** cubes.

Five of the cubes are grey. The other four cubes are white.

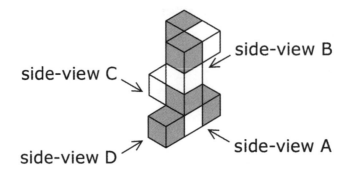

side-view B

side-view C

side-view A

side-view D

a   The drawings below show the four side-views of the model.

Which side-view does each drawing show?

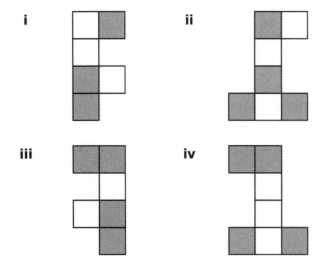

i

ii

iii

iv

1 mark

continued.

b   Copy and complete the **top-view** of the model by shading the squares which are **grey**.

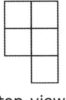

top-view

1 mark

c   Imagine you turn the model **upside down**.

What will the new top-view of the model look like?

Copy and complete the **new top-view** of the model by shading the squares which are **grey**.

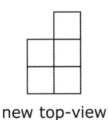

new top-view

1 mark

Here are some extracts from a newspaper report.

Seats at new Wembley will cost up to £6000 per year

Demolition work on the famous twin towers, the most recognisable feature of Wembley stadium, started yesterday but the new venue that will take its place is set to be the most expensive sports arena in the world with seats costing up to £6000 per year and executive boxes up to £200 000.

It is the cost of the premium seats and executive boxes that will most shock football fans.

There will be five different categories of premium seats which have to be purchased for a minimum of 10 years. The cheapest premium seat will cost £1300 per year... with the most expensive costing £6100 per year.

In addition to the yearly payments, firms and individuals buying the premium seats will also have to make a one-off payment to

Either:

Find a magazine or newspaper article about an issue that you could investigate statistically.

Or:

Use the article above.

Using your chosen article:
- Specify a problem you could investigate on this topic.
- Decide on a hypothesis you could test.
- Decide on the data you would need to collect, and where you would get it.

Write an explanation of the decisions you make.

Peter is investigating people's television viewing habits.
Here is part of a questionnaire he designs:

Name _____ Age_____ Address _____

1 How much television do you watch?

☐ At lot ☐ Not much ☐ Very little ☐ None

2 What is your favourite type of programme?

3 Do you think there is too much sport on television?

4 How many televisions do you have in your house?

5 What channel do you watch the most?

6 'There are too many boring programmes about things that only adults would be interested in. There should be more programmes for young people. After all, we are going to be the ones paying for TV licences in a few years' time, so they should give us what we want.'

Do you:

STRONGLY AGREE / AGREE / PARTLY AGREE / DON'T UNDERSTAND THE QUESTION.
(Cross out the parts that do not apply.)

There are some parts of the questionnaire that could be better!

◆ Design an improved version of the questionnaire, and explain why you made the changes you decided on.

◆ Write some 'Top Tips for Designing a Quality Questionnaire'. This should be a list of hints and suggestions to help people design a questionnaire that is clear and easy to use.

> **Remember:**
>
> To use an assumed mean:
> - ◆ assume a convenient value for the mean
> - ◆ subtract it from each piece of data
> - ◆ find the mean of the new values
> - ◆ add the assumed mean back on.

Work out the mean of each of these sets of data, using an appropriate assumed mean. You should show all of your working.

1 Number of ChocoDrops in 10 different packets

23, 23, 25, 24, 22,
22, 24, 25, 23, 22

2 Diameter (in mm) of 20 broomstick handles

26.0, 25.7, 25.7, 25.9, 25.8,
25.9, 26.0, 25.6, 25.8, 25.6,
25.3, 25.9, 26.1, 25.3, 26.0,
25.7, 25.9, 25.5, 25.4, 25.6

3 Weight (in grams) of 20 eggs

65, 71, 66, 63, 76, 73, 62, 63,
69, 69, 61, 66, 64, 69, 69, 68,
63, 61, 62, 70

4 Finishing times (hours:minutes) for 12 marathon runners

3:12, 2:55, 3:08, 3:26, 2:47,
3:15, 3:18, 3:05, 2:49, 3:17,
3:22, 3:10

Be particularly careful with the last one – the times are not given as decimals!

Here are two hypotheses from statistical projects, and scatter diagrams showing data that was collected to test each hypothesis.

Write a conclusion for each project.

Hypothesis: Larger countries have bigger populations.

The scatter diagram shows data for 10 different countries.

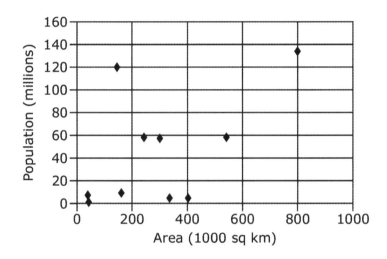

Hypothesis: People's marks in two maths tests were similar.

The scatter diagram shows two sets of test results for a class.

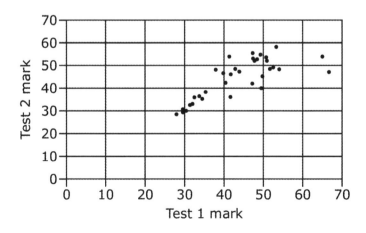

A newspaper predicts what the ages of secondary school teachers will be in six years' time.

They print this chart.

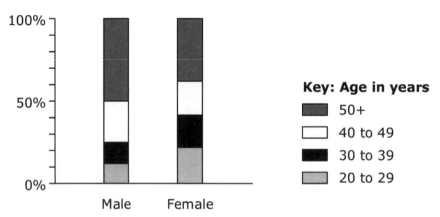

Key: Age in years
- 50+
- 40 to 49
- 30 to 39
- 20 to 29

a The chart shows **24%** of male teachers will be aged 40 to 49

About what percentage of female teachers will be aged 40 to 49?

1 mark

b About what percentage of **female** teachers will be aged 50+?

1 mark

c The newspaper predicts there will be about **20 000** male teachers aged 40 to 49.

Estimate the number of male teachers that will be aged 50+.

1 mark

d Assume the total number of male teachers will be about the same as the total number of female teachers.

Use the chart to decide which statement is correct.
Write down the correct statement.

i Generally, male teachers will tend to be younger than female teachers.

ii Generally, female teachers will tend to be younger than male teachers.

Explain how you used the chart to decide. *1 mark*

A school has 5 Year groups.
80 pupils from the school took part in a sponsored swim.
Lara drew this graph:

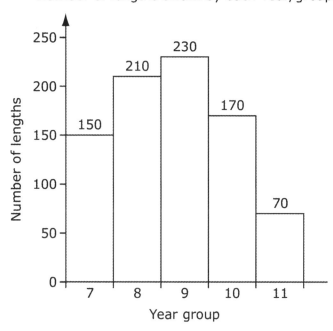

Number of lengths swum by each Year/group

a Look at the graph.
Did **Year 10** have **fewer** pupils taking part in the swim
than **Year 7**?
Choose from:

Yes   No   Cannot tell

Explain your answer.   *1 mark*

b Use the graph to work out the mean number of lengths swum
by each of the 80 pupils.

Show your working.   *2 marks*

This game is based on noughts and crosses.

It is a game for one player.

You will need a coin, a piece of blank paper and a pencil.

The aim of the game is to get a line of three crosses in a row,

perhaps like this: or like this: or like this:

1   Draw a blank 3x3 grid.

2   Throw the coin.

If it lands heads, insert O on any one of the nine squares.

If it lands tails, insert X.

3

Repeat step 2 until:

♦   you get a straight line of three crosses

⇒   you have won

♦   you get a straight line of three noughts

⇒   you have lost

♦   you fill the grid, with no clear line

⇒   it is a draw

Repeat the game a few times.

Focus on the best strategies to use.

Write a short report on your results, and answer the question:

♦   Which are you most likely to do: win, draw or lose?

Explain your answer.

The aim of this activity is to find the probability of getting 'heads' when you throw an ordinary coin.

You know that the theoretical probability of getting heads is 0.5...

... and the theoretical probability of getting tails is 0.5.

1　**a** Throw a coin 10 times and record your results.

　　b Use your results to estimate the probability of getting heads.

2　Now throw the coin another 10 times, making 20 results altogether.

　　Use the complete set of 20 results to make a new estimate of the probability of getting heads.

3　Now throw the coin 30 more times.
　　Use the complete set of 50 results to make another estimate for the probability of getting heads.

4　Record the results for 50 more tosses of the coin.

　　Make a final estimate for the probability of getting heads, using the complete set of 100 results.

5　Write about your results.

　　You would expect the experimental probability to get closer to the theoretical probability as the number of trials increases.

　　Is this what actually happened?

On a particular stretch of road there are three sets of traffic lights.

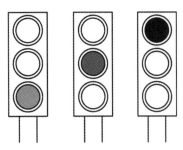

For each set of lights, at any particular time:

The probability of the lights being green is $\frac{1}{2}$

The probability of the lights being amber is $\frac{1}{6}$

The probability of the lights being red is $\frac{1}{3}$.

Laila knows that if all three sets of lights are green, she can cycle home in five minutes. In a typical week, Laila cycles along the stretch of road 10 times.

Use a dice to perform a simulation of the traffic lights.

Hence estimate the number of times in a week that Laila can expect to cycle home in five minutes.

a A spinner has **eight** equal sections.

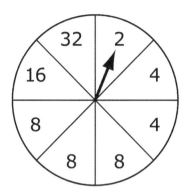

What is the probability of scoring **4** on the spinner?

1 mark

What is the probability of scoring an **even** number
on the spinner?

1 mark

b A different spinner has six equal sections and **six numbers**.

On this spinner, the probability of scoring an **even** number is $\frac{2}{3}$.

The probability of scoring **4** is $\frac{1}{3}$.

Write down what numbers could be on this spinner.

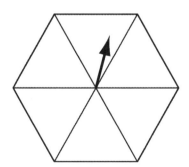

2 marks

I have two fair dice.

Each of the dice is numbered 1 to 6.

a   The probability that I will throw **double 6** (both dice showing number 6) is:

$$\frac{1}{36}$$

What is the probability that I will **not** throw double 6?   *1 mark*

b   I throw both dice and get double 6.
Then I throw the dice again.

Write down the statement that describes the probability that I will throw **double 6** this time:

◆   less than $\frac{1}{36}$

◆ $\frac{1}{36}$

◆   more than $\frac{1}{36}$

Explain your answer.   *1 mark*

I start again and throw both dice.

c   What is the probability that I will throw **double 3** (both dice showing 3)?

   1 mark

d   What is the probability that I will throw a double?
(It could be double 1 or double 2 or any other double.)

   1 mark